Space Elevators: The Green Road to Space

Editor: Jerry Eddy, Ph.D.

Peter Swan, Ph.D.
Cathy Swan, Ph.D.
Paul Phister, Ph.D.
David Dotson, Ph.D.
Joshua Bernard-Cooper
Bert Molloy

Prepared for the
International Space Elevator Consortium

April 2021

Space Elevators: The Green Road to Space

Published by Lulu.com

info@isec.org

978-1-6780-8710-4

Cover Illustrations:
Nixene Publishing, with Image sourced from Pixabay

Printed in the United States of America

Preface

This 18-month study has shown some remarkable results during a time of transition inside the space arena. As the needs grow for more and more assets in space (at GEO, the Moon and Mars) the realization grows that a robust permanent transportation infrastructure is desperately needed. Since many are dreaming of permanent colonies in space and numerous countries are looking for ways to satisfy their present and future power needs it becomes apparent that huge amounts of mass will be needed in space if these dreams are to be realized. Can it all be done with rocket launches? We have three messages that result from our study. We believe they should be considered in any discussions on "how to" fulfill these space dreams. Essentially, we (ISEC) must insert the "positive environmental impact" and Space Elevators into all aspects of future space endeavors.

- Message ONE: Our visions match yours! We are building the Green Road to Space in response to so many customer needs. A key aspect is that the tether climber is propelled by electricity as it raises off the surface of the Earth enabling a zero-carbon footprint during operations.
- Message TWO: Our strategy is to propose a Dual Space Access Architecture: First, rockets will be better, less expensive, more robust and reusable in the future; but, their payloads will still be restricted by the rocket equation. As such, create a joint venture, rockets moving the valuable assets to any orbit rapidly through radiation belts now and in the near future; and Space Elevators transporting huge cargoes to desired destinations. The second component of the Dual Architecture will be Space Elevators as permanent infrastructures. They will come in to play in the late 2030's with an ability to move massive tonnage to customers' destinations; daily, routinely, and inexpensively while doing it Earth-friendly. This Dual Access combination of future transportation elements will leverage both sets of strengths and enable greater mission successes.
- Message THREE: The authors are ready to initiate a Space Elevator Developmental Program: The tether material (Single Crystal Graphene) is in the laboratory now and will be available in time for development. The current belief is that the major segments (Earth Port, Galactic Harbour and climbers) can be started now so that they are ready when the tether material becomes available. But the key is that the development program can begin now.

Acknowledgements

Having a vision of the future is the first step that leads to programs developing dreams. This process has been around forever; however, for the future of off planet movement, this dreaming has been accelerating over the last 200 years. The last 63 years has shown on how hard it is to accomplish space-based visions. Now we have visions from so many - NASA and ESA with the Artemis Program, Mr. Bezos and Mr. Musk hope to move off planet in a big way; and recently, three countries have orbited/landed on Mars. As for this report, first and foremost, each author needs to be acknowledged for the supreme effort they gave in doing the extensive research as well as the writing efforts they contributed in their respective chapters. We would also like to thank the reviewers for pointing out the many inconsistences from one chapter to the next. Vern Hall, Paul Phister, David Raitt and Dennis Wright improved this document with their remarkable insights.

Thanks must also go to the members of the International Space Elevator Consortium, and other Space Elevator enthusiasts who have been dreaming "big" for years now. The phenomenal work accomplished over the last ten years has allowed the Space Elevator body of knowledge to increase exponentially. Each of us knows that when the Space Elevator is operational, movement off-planet proceeds robustly. We know that we will have an impact on the future and move humanity with hope. This Massive Transformative "Moonshot" is rewarding to work on and is being accomplished because we believe. Well done Space Elevator team!

Dedication to Joyce:

As the main editor of the report, I would like to dedicate this work to my wife of 58 years, Joyce Porter Eddy who passed away from pancreatic cancer December 18, 2020. I personally wish to thank Dr Peter Swan for relieving me of much of my responsibility, while I was my wife's principal care giver.

Jerry Eddy

Executive Summary

The year of 2020 had many negative aspects from fires, pestilence, hurricanes and pandemics. The beautiful thing is that it also had so many nations coming together in the vision of moving off-planet. The summer flotilla of vehicles going to Mars was remarkable and the joining of countries to go to the Moon shows an initiative so huge it is motivational. The question then became, with this new movement off-planet, how shall the Space Elevator enthusiasts respond? All this action reinforces the critical nature of the Space Elevator as participants in the future with the inherent strengths of being a permanent transportation infrastructure with a zero-carbon footprint. If everyone wants to have their citizens living on the Moon or Mars), a massive movement of equipment and supplies will be needed. Space Elevators are the answer! The Space Elevator community's vision should be, that it can support this historic movement and ensure its success. Space Elevators CAN move millions of tonnes of cargo - no-one else can with a beneficial environmental approach and timely delivery to multiple destinations. Our new vision is:

> *Space Elevators are the Green Road to Space - they enable humanity's*
> *most important missions by moving massive tonnage to GEO and beyond.*
> *They accomplish this safely, routinely, inexpensively, daily and they are*
> *environmentally neutral.*

This study report will show how the Space Elevator is a Massive Green Machine and should be called the "Green Road to Space." In addition, the study report will show how the Space Elevator enables missions that cannot reasonably be accomplished with rockets and thus can help improve the human condition on Earth.

The study reports on the "green missions" of Space Solar Power, Sun Moon L-1 Solar Shade, and permanent disposal of high-level nuclear waste. In addition, it assesses the environmental impact from development and operations of Space Elevators. One of the main conclusions is that the movement off-planet demands the initiation of a Dual Space Access Architecture where future rockets and Space Elevators are complementary, compatible and not competitive.

Indeed, others are ready to leap into the off-planet movement. However, Space Elevators, as a part of the Dual Space Access Architecture, have tremendous strengths that have not yet been included in their strategies for going to the Moon and beyond. This new movement off-planet should include the Space Elevator's ability to:

- Depart the Apex Anchor at great velocity (7.76 km/sec)
- Support interplanetary missions (Fast Transit to Mars 61 days)
- Supply massive daily payloads (170,000 tonnes per year)
- Create entrepreneurial enterprises at GEO and the Galactic Harbours.

- Enable new environmentally significant missions (Space Solar Power, Solar Shades, hi-level nuclear waste disposal, etc.)
- Enable carbon negative operations for delivery to orbit
- Exit the gravity well using solar powered tether climbers and not rocket fuel and accomplish this daily, routinely, inexpensively and with carbon negativity

Indeed, the implementation of the Dual Space access Architecture should follow:

> *Rockets to open up the Moon and Mars with Space Elevators to supply and grow the colonies. In addition, Space Elevators will enable Green Missions such as, Space Solar Power and L-1 Sun Shades. This approach is compatible and complementary with future rockets while leveraging the strengths of both inside a Dual Space Access Architecture.*

Table of Contents

Chapter 1 - Introduction to "Green" Assessment Trade Study

1.0 Introduction:

The Space Elevator will be the transportation story of the 21st century. Reliable, routine, safe, environmentally friendly and efficient access to space is close at hand. The Space Elevator is the backbone of the Galactic Harbour, and will be an essential part of the global and interplanetary transportation infrastructure. This study evaluated the Space Elevator as a significant component of humanity's movement into space. The study group looked at the needs for massive movement of cargo to support significant missions of the future and the impacts of these activities on the Earth's environment. This led to the realization that there were two components to this question: (1) what could Space Elevators contribute to solve some of our serious problems, and (2) what would be the environmental impacts of construction and operations of Space Elevators. These two questions lead to assessing Space Elevator environmental impact and evaluating the contributions of other programs that can be enabled by Space Elevators. As the study team addressed these "big questions," especially as humanity is moving off-planet, Space Elevators, along with future rockets, must be a major part of the discussions. The combination of these two significant components becomes a Dual Space Access Architecture (DSAA) (see chapter 2).

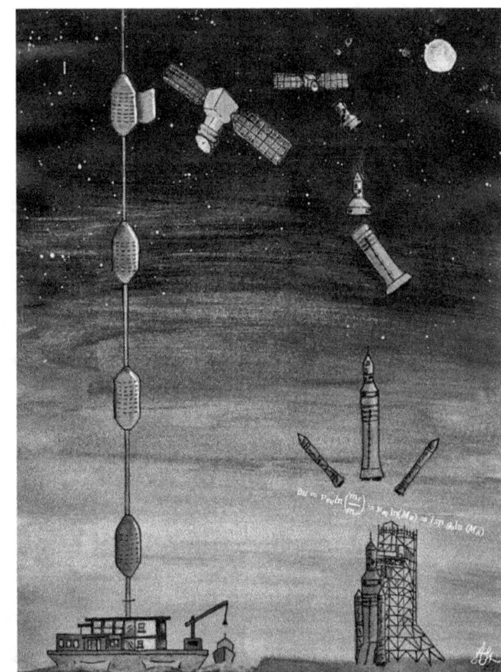

Figure 1: Dual Space Access Architecture (Stanton image)

A permanent transportation infrastructure would have characteristics of daily, routine, inexpensive, environmentally friendly, and massive tonnage of payloads to GEO and beyond. Release from several locations on the Space Elevator will allow high speed launches to other bodies in our solar system. Our previous study "Space Elevators are the Transportation Story of the Twenty First Century" [Swan, 2020a] showed that Space Elevators allow massive amounts of material to be moved routinely to GEO (upwards of 80 tonnes daily). This report will show that this massive movement by Space Elevators allows the additional benefit of solving many of Earth's environmental problems, most importantly global warming. This study will start the discussions by showing the additional benefits of Space Elevators being "Massive Green Machines." The current vision is:

> *"Space Elevators are the Green Road to Space - they enable humanity's most important missions by moving massive tonnage to GEO and beyond."*

The reality is that when humanity decides to go off-planet, there will be a tremendous need for logistical support, movement of manufactured goods as well as transporting people [especially at low cost and routine/daily]. The question on the table is: how can the strengths of Space Elevators enable missions of all types, while having minimal or no environmental effect on our planet? We believe that not only can Space Elevators do this, but they also enable activities in space that will improve Earth's environment. This study shows the environmental effects of building and daily use of Space Elevators and the resulting activities which can be accomplished to improve the Earth's environment.

The study team decided to make the main thrust showing the "Greening of the Earth" resulting from the Space Based Solar Power Program (SSP). This study group looked at several space missions that would improve the health of the Earth's environment and asked the question: "Can the Space Elevator enable missions in space that would improve the human sustainable condition?" The number one concept with the most potential for "good" was the Space Solar Power system. Prior to this study activity, the beneficial impact of a SSP program was explained by Dr. John Mankins as he stated: "an extensive SSP program can stop global warming and possibly reverse it." Previously, the SSP experts recognized two significant developmental factors: (1) the massive program, with acres of solar cells at GEO would require five million tonnes [Mankins, 2019c] delivered to approximately 36,000 km altitude, and (2) previous SSP design engineers had determined that the massive need was so great, they would have a better approach by building the massive components (solar arrays and antennas) on the Moon from regolith.

 International Space Elevator Consortium study group's conclusion is that the implementation of a SSP project can be done with A) Space Elevators or B) a large number (towards tens of thousands) of rocket launches, or C) a Dual Space Access Architecture - combining the best of both methods. This report will show that SSP, on a large scale, can only be accomplished in a green manner by Space Elevators moving large tonnage from Earth's surface to GEO. Thus, a faster and cheaper SSP approach can be realized for the betterment for all mankind.

In addition to SSP, the study group believes that many of the space activities that are being planned can be done much easier, less expensively, and greener with Space Elevators. The study team also considered the High-Level Nuclear Waste (H-L-W) problem and how Space Elevators can be used to remove H-L-W from the Earth and disposed of safely (orbit close to the Sun). This is a consideration that up till now has not been seriously been considered because of the slight potential of serious nuclear accidents with rockets. Another "Greening Technology" Program considered is the placing of large sun shields or shades in space to reduce solar energy from reaching Earth and thus reduce global warming.

The ISEC acknowledges that rockets will always be needed to convey people and materials to space that need to be placed there quickly. It is the combination of future rocket and Space Elevator capabilities into a Dual Space Access Architecture that seems to have a remarkable future. The following "thrusts" show how Space Elevators become a Big Green Machine:

Potential Beneficial Impact of Space Elevator:
- Reducing the number of fossil fuel burning plants providing energy (100s of coal plants) by the delivery of energy from orbit to anywhere all the time (SSP).
- Providing safe disposal of High-Level Nuclear Waste.
- Reducing the energy from the Sun that reaches the Earth's Atmosphere, thus reducing global warming.
- Reducing the number of launches (such as to support humanity's movement off-planet) will decrease pollution significantly.
- Providing safe, reliable, routine, daily, environmentally friendly, and inexpensive transportation infrastructure to move massive tonnage to GEO and beyond, specifically the Moon and Mars.

Problem Definition: As humanity has decided to move off-planet, the frequency of lift-offs will increase significantly in order to transport massive amounts of materials and people towards GEO, Moon and Mars. The effects upon the Earth's environment will ramp up tremendously if we only use rockets to access space in our future. Going from 100 launches per year, with damaging effluences burning in the atmosphere during thrust, to thousands of launches to support a movement to the Moon and/or Mars, will excerbate the modest environmental impact of today. This environmental impact from rocket launches has been assessed; however, it is essentially overlooked today, because rockets are the only access to strategic missions in space. This study report will look at the environmental benefits of Space Elevators and show that these beneficial environmental effects should be recognized leading to a critical demand for the Dual Space Access Architecture, future rockets and Space Elevators working cooperatively. A quick summary of our study results show:

Space Elevators are significantly carbon negative and contribute to the betterment of the Earth's atmosphere from the missions enabled.
- Space Elevators will enable Massive Green Technology missions improving the Earth and its environment. (Space Based Solar Power being the principal example)
- Space Elevator operations will have zero environmental footprints.
- Rockets are essential and the environmental impact must be understood

Background: The envisioned future missions to GEO for space solar power and colonies on Mars and the Moon are of incredible size and scope, require a huge amount of logistical support. If accomplished by rockets alone, this will require a large number of rocket launches with a possible negative impact on the Earth's environment. This report shows that operational Space Elevators would have little or no environmental impact and that a number of environmental problems could possibly be solved with their usage. This study investigated the environmental effects of the construction and operation of a Space Elevator. It will also consider what possible Earth environmental problems could be alleviated or even solved by the use of Space Elevators. This architecture would leverage the strengths of each major component - rockets and Space Elevators.

Approach: This study approached the future with a remarkable concept: Dual Space Access Architecture enabling massive liftoff capacities will ensure a remarkable movement off-planet. The initial chapters set the stage and expand the background needed to capture this concept. Individual chapters lay out the explanation of the environmental impacts from Space Elevators (chapter 6) and show the environmental benefits of missions that can only be accomplished by the massive movement of satellite segments via the Space Elevator (chapter 3, 4, & 5).

1.1 Developmental Status of Space Elevators and Galactic Harbours:

Recently, a visitor to ISEC's conference and their website (isec.org) was quoted as saying, "You have a remarkable body of knowledge." He was referring to the efforts of many scientists, engineers, and project/program professionals over the last eight to ten years. The leap in quality and currency shows that the Space Elevator is indeed twenty years beyond Dr. Edwards' breakthrough accomplishment saying "it can be done." What are amazing are the conclusions from this body of knowledge:

- Space Elevators are ready to have a large-scale developmental program started.
- The tether material has been produced in the laboratory for the needed strength (150 GPA) and continuous length (1 meter per minute production) (note; not both capabilities at once - yet). One of several 2D materials, single crystal graphene or hexagonal boron nitride, will be ready for the development team.
- Space Elevators enable Missions off-planet by robust cargo movement.
- Space Elevators are environmentally friendly in operations and enable Space Based Solar Power to eliminate hundreds of coal burning plants.

Justification for the statement - "Space Elevators are ready to have a developmental program started" - has arisen from a massive amount of Space Elevator research over the last 18 years. Several of these following documents explain the rationale and "show the numbers:"

- "Today's Space Elevator" [Swan 2019a] lays out the status of the developmental program as of summer of 2019.
- "The Road to the Space Elevator Era" [Swan 2019b] shows an excellent summary from four years of analysis at the International Academy of Astronautics with identification of the main Space Elevator technologies and their readiness for flight.
- "The Space Elevator is the Transportation Story of the 21st Century," [Swan 2020a] shows an excellent explanation of the near term "customer demands" across several missions (Mars Colony, Lunar Village, and SSP).

1.2 Introduction to Assessment Approach:

This study has been authorized by ISEC to illustrate the tremendous strengths of the Space Elevator and what environmental effects, both negative and particularly positive, it can have.

The concept of a net assessment of a future problem is: To provide executive level management with an appraisal of the state of affairs that affect the character and success of the total enterprise. This report will look at the issues related to the future use of massive numbers of rocket launches and compare them to the issues of massive movement of cargo on a Space Elevator. This net assessment will look at three factors and compare the analysis for future architecture recommendations. The analysis must deal with a three factor interrelated assessment including:

1) Forcing Function of how much mass (people and cargo) is required for future missions to which orbit (Table 1.1 Customer Demand).

Table 1.1: Explanation of Demand Pull from Customers [Swan 2014]

Demand in Metric Tons	2031	2035	2040	2045
Space Solar Power	40,000	70,000	100,000	130,000
Nuclear Materials Disposal	12,000	18,000	24,000	30,000
Asteroid Mining	1,000	2,000	3,000	5,000
Interplanetary Flights	100	200	300	350
Innovative Missions to GEO	347	365	389	400
Colonization of Solar System	50	200	1,000	5,000
Marketing & Advertising	15	30	50	100
Sun Shades at L-1	5,000	10,000	5,000	3,000
Current GEO satellites + LEOs	347	365	389	400
Total Metric Tons per Year	58,859	101,160	134,128	174,250

2) Impact Function of how much environmental damage results from Space Elevators satisfying the customer demands - see next chart showing # flights vs liftoffs.

Table 1.2: Number of Rocket Launches vs. Galactic Harbour Lift-Offs [Swan 2020]

Type	Lift Average (tonnes)	Number per year	Total Mass to Interplanetary (tonnes) per year
Individual Heavy Launch Vehicles Current	10 per launch	87 (average last 5 years)	870 tonnes if all went to single mission
Individual Heavy Launch Vehicles - SpaceX Starship (estimate 2030)	100 per launch to LEO - 21 towards Mars	100 times per year	2,100 tonnes
Galactic Harbour Transportation Infrastructure at Initial Operational Capability (estimate 2040)	6 tethers x 14 tonnes = 84 per day	every day towards Mars and Moon	84 x 365 = 30,660 tonnes
Galactic Harbour Transportation Infrastructure at Full Operational Capability (estimate 2050)	6 tethers x 79 tonnes = 474 per day	every day towards Mars and Moon	474 x 365 = 173,010 tonnes

3) Impact Function of how much benefits comes from new missions:

Table 1.3: Mission Benefits

Program	Mission Benefits	Mass to Destination (tonnes)
Space Solar Power	Eliminates 100's of Coal plants	5,000,000
Sunshade at Sun-Earth L-1 location	Reduces radiation to Earth by 2% of total	11,000,000
Permanent disposal of high-level nuclear waste	Permanently eliminates high level nuclear waste	Millions of tonnes

The approach in this study followed these steps:

Step One: The most important part of the study was to consider all the ways a Space Elevator could be used to solve (enable) environmental problems that now plague the Earth. Some of these future programs (not achievable today through rockets) are:

- SSP: As suggested by O'Neill, any sizable space colony's main purpose would be to construct large solar panels to capture the sun's rays and beam the energy to Earth using microwaves [O'Neill, 1974]. It was thought that this would solve the problems created by burning fossil fuels and thus give Earth an alternate power source as fossil fuels become depleted. Now the discussions are around using Lunar regolith as the major component of solar arrays built on the Moon and delivered to the GEO ring. Space Elevators can lift the required mass from the Earth and skip either extraneous missions for Lunar or orbiting colonies. (Chapter 3)
- High Level Nuclear Waste Removal:" Could massive nuclear waste be disposed of economically and routinely in outer space? This will also be considered along with

how it should be done, projection into the sun, Jupiter or in some other way. Solving the nuclear waste problem would encourage more nations to rely on nuclear power instead of fossil fuel fired power plants. (Chapter 4)
- Solar energy blockage at the Earth Sun L-1 orbital location: Blockage of a percentage of solar energy before it reaches Earth's atmosphere could help in lowering the global temperature. This would require massive amounts of material delivered across a wide area in the L-1 location. (Chapter 5)
- Manufacturing that is polluting or dangerous on the Earth could be accomplished safely at GEO, with minimal expense for movement of materials safely to GEO and finished products returned to Earth or sent on to Moon or Mars colonies. (discussed, but not presented)
- Planetary Defense: A new concept surfaced where the Apex Anchors became Storage locations for massive defensive equipment; to include solid booster rockets and deflection techniques. (discussed, but not presented)

Step Two: Summary of Environmental Assessment of Space Elevators.
- The construction and operations of a Space Elevator, considering the construction of the tether, ocean platform, climbers and transporting of materials to the construction site. (Chapter 6)

1.3 An insight:

A recent article entitled "Why addressing the environmental crisis should be the Space industry's Top Priority," describes "how can we give meaning to space missions in the context of global environmental crisis?" [Miraus, 2020] The thought is that improving the approach to space, and access to space, can (and should) include discussions about the environmental damage. This is a timely and critical discussion that should be initiated.

"For a green space industry: A completely different model of society needs to be found and implemented to respect the planetary boundaries while staying above social thresholds. In this context, it is legitimate to question the practices of the space industry and its role in achieving this goal. We, space people, can decide to include in our efforts and jobs a greater environmental ambition. We can transform our industry to be dedicated to monitor and mitigate climate change, to efficiently manage the increasing number of disasters and crises resulting from its effects, and to help the transition towards a low-carbon society, all while avoiding development of space applications on Earth that are polluting, that favor consumerism, or that accelerate the destruction of the environment. We can make our industry itself more sustainable. Ecology should be placed at the heart of space activities". [Miraux, 2020]

1.4 Conclusions:

The future must move to the Dual Space Access Architecture concept with Space Elevators moving massive tonnage while the Galactic Harbour encourages and develops space enterprises along their vertical "train tracks." Rockets, as complementary to Space Elevators, have a significant role for delivery of people and valuable cargo and have their place when delivering payloads to LEO or MEO. It is basically economical with minimal impact to the Earth's environment. However, when venturing beyond LEO or MEO, to GEO and other planets it simply is not feasible to "build a bigger" rocket that has to make tens of thousands of launches to deliver the required payloads. Utilizing the capabilities of Space Elevators, coupled with future rockets, to create a "Dual Space Access Architecture" is the most efficient, cost effective way to deliver payloads outside Earth's neighborhood.

In addition, this study report will show the potential benefits of Space Elevators when reaching for the stars, especially with respect to their zero-carbon footprint.

Table 1.4:   Potential Beneficial Impacts of Space Elevator:

Approach	Effect
Enabling Space Solar Power	Reducing the number of fossil fuel burning plants providing energy (100s of coal plants) by using the delivery of energy from orbit to anywhere all the time
Zero (or negative) carbon footprint to deliver to space	Daily operations, at zero (or negative) carbon footprint, reduces the environmental impact of the expected massive movement to space.
Enable Appropriate Solar Shade at L-1	Reducing the energy from the Sun that reaches the Earth's Atmosphere, thus reducing global warming.
Reduce Buring of fuel in Atmosphere	Replacing number of rocket launches (such as to support humanity's movement off planet) will decrease pollution significantly.
Environmentally Friendly Space Infrastructure	Provides safe, reliable, routine, daily, environmentally friendly, and inexpensive transportation infrastructure to move massive tonnage to GEO and beyond, specifically the Moon and Mars.
Enable Permanent Disposal of High Level Nuclear Waste	Deposits Nuclear Waste in Small Solar Orbit which provides safe and long term storage of High-Level Nuclear Waste.

A Dual Space Access Architecture combining rocket and space elevator strengths results in tremendous advantages in the "greening of the Earth." The first is an insight: rockets are here now and already robust, resulting in rapid transit through radiation belts with people. The second is that all the robotic movement of mass in the future (cargo, habitats, air, water, etc) would be moved safely, routinely, daily, environmentally friendly, and inexpensively by Space Elevators. This separation of delivery approaches will greatly enhance missions. As customer demand for huge masses matures to support near term missions such as Space Based Solar Power (five million tonnes to GEO) and a Mars Colony (one million tonnes to Mars), the value of Space Elevators becomes obvious. When the Space Elevator delivers 75% of the mass needed for critical missions, the savings in cost, time and environmental impact will make us ask: Why not sooner?

This study will start the future discussions by showing the additional benefits of Space Elevators being defined as "Massive Green Machines." The current vision is:

"Space Elevators are the Green Road to Space - they enable humanity's most important missions by moving massive tonnage to GEO and beyond."

In point of fact, the operations of Space Elevators and Galactic Harbours will be carbon negative. Several of the concepts developed above can be considered key elements establishing the reality that Space Elevators will make the Earth greener. This net assessment trade study conducted by ISEC showed that:

Space Elevators and Galactic Harbours are Big Green Machines designed to improve the Earth's environment through two significant contributions. The first is the remarkable "zero-emission" lift of cargo to space - reducing environmental impacts from rocket launches. The second is the ability to deploy massive systems to GEO and beyond that eliminate rocket launches by becoming a partner in Dual Space Access Architecture.

1.5 Study Report Breakout:

Chapter 1   Introduction to Net Assessment Trade Study
Chapter 2   Galactic Harbour 2020 Vision
Chapter 3   Space Elevator Enabled Environmental Benefits - Space Solar Polar:
Chapter 4   Space Elevator Enabled Environmental Benefits - High Level Nuclear Waste Disposal:
Chapter 5   Space Elevator Enabled Environmental Benefits - L-1 Solar Shade
Chapter 6   Galactic Harbour Environmental Impact
Chapter 7   Environmental Impact Assessment of Dual Space Access Architecture
Chapter 8   Conclusions
Chapter 9   Recommendations

References

Appendices:
A - International Space Elevator Consortium
B - Avoiding the Rocket Equation,
C - Body of Knowledge
D - Studies (ISEC, IAA, Obayashi)
E - Hi-Level Nuclear Waste Appendix
F - Space Solar Power Backup

Chapter 2 - Galactic Harbour 2021 Vision

2.1 Introduction:

This chapter establishes the background information to place Space Elevators and Galactic Harbours inside the future of the movement off-planet. The authors believe a robust movement beyond low Earth orbit requires permanent infrastructures designed to move cargo inexpensively, safely, routinely, daily, in an Earth-friendly way with massive tonnage capability. The Space Elevator permanent transportation infrastructure, inside Galactic Harbours, will enable humanity to reach these goals during the first half of this century. At this time, humanity needs to expand its vision and continue the growth of the species by conducting scientific investigations, human exploration, commercial developments and then human habitats and colonies on the Moon and Mars.

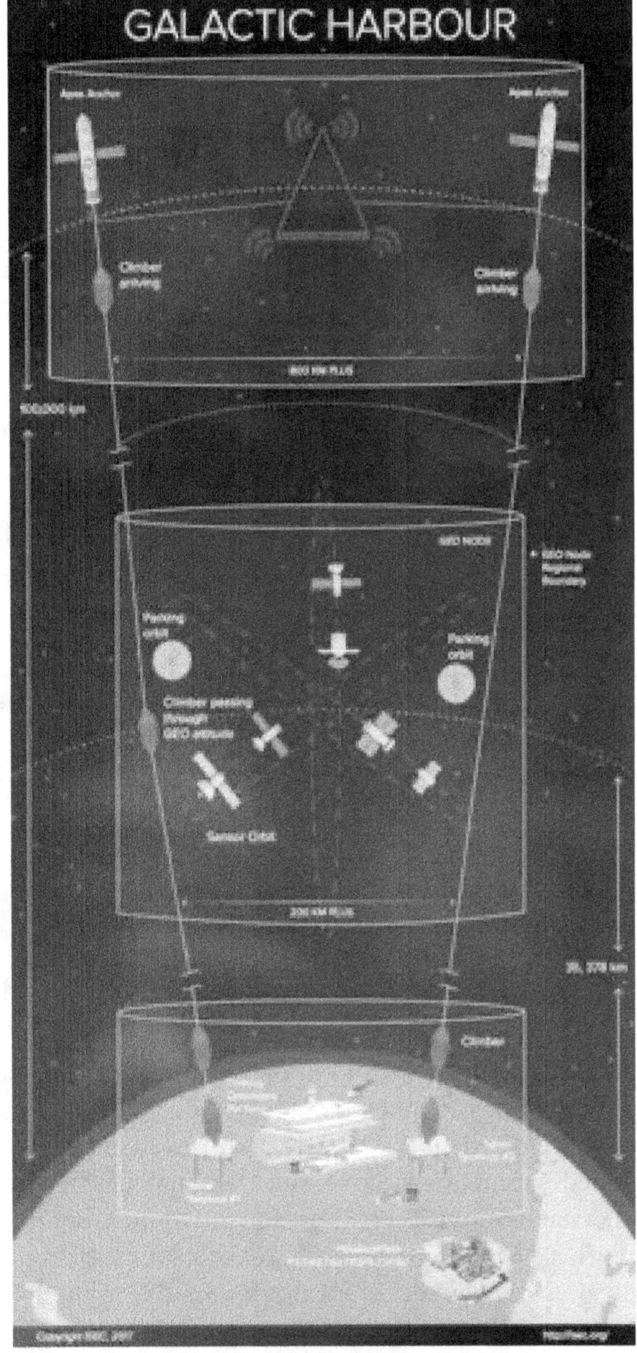

Figure 2.1, Galactic Harbour

As shown in the recent year-long ISEC study report [Swan 2020a], the concept development supporting interplanetary movement led to developing a vision of future Space Elevators and Galactic Harbours. But first, what is a Galactic Harbour?

The Galactic Harbour: [Swan, 2020a] Galactic Harbours are the unification of Transportation and Enterprise. From an engineering aspect, Galactic Harbours are a combination of Space Elevator Transportation Systems and Space Elevator Enterprise Systems. A Galactic Harbour will be the volume encompassing the Earth Port while stretching up in a cylindrical shape to include two Space Elevator tethers outwards beyond GEO to the Apex Anchor.

Customer product/payloads will enter the Galactic Harbour at an Earth Port and exit someplace up the tether. There will be tremendous enterprise development in the GEO Region such as: spacecraft assembly, refueling operational satellites, solar power collection and, of course, businesses will emerge supporting flight operations from the Apex Anchor to interplanetary destinations. From an operational aspect, The Transportation System is the "main channel" in the Galactic Harbour moving cargo from the Earth Port to transportation locations within the Harbour - i.e. the GEO Region and the Apex Anchor Region. New businesses at GEO, the Apex, and at the Earth Port are the Space Elevator Enterprise system. When the Elevator becomes operational, it will service all of these enterprises.

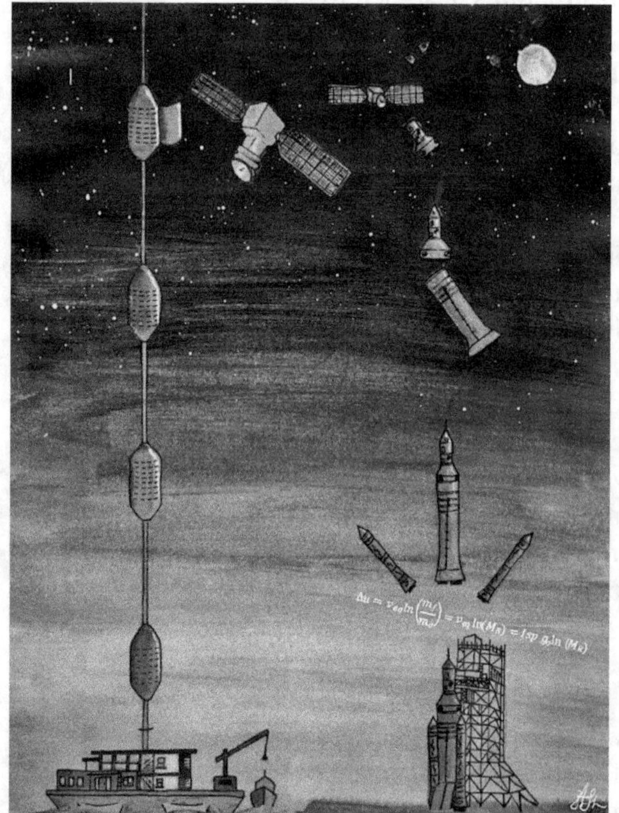

Figure 2.2, Dual Space Access Architecture (A Stanton Image)

2.2 Dual Space Access Architecture:

With this concept of Galactic Harbours comes the recognition that movement off-planet will require complementary capabilities, such as rocket portals and Galactic Harbour infrastructures, each with their own strengths and short-falls. While editing the latest ISEC year-long study report, the authors recognized some powerful truths. One of the biggest is that Space Elevators will stand up strong next to rockets and help enable movement off-planet. When the authors look at the Moon and dream of spaceflight, they forget how extremely difficult it was to accomplish - both in energy and design complexity. Tsiolkovsky's rocket equation consumes so much mass to achieve orbit that, historically, we have been restricted as to what we can deliver. Now that we have decided to go to the Moon, and on to Mars, in a combined international, governmental, and commercial effort of great magnitude, we need to expand our vision of 'how to.' It would seem that the establishment of a more robust infrastructure with reusable rockets and permanent Space Elevators must be developed. ISEC will develop and present the strengths and weaknesses of the two major components of this combined architecture with the purpose of placing mission equipment and people where they need to go and when they need to be there. The multiple destinations, complexity of orbits, magnitude of each transition to orbit, and infrequent launches currently means that the difficulty of fulfilling the dreams of the many is a monumental "reach." Expanding space access architectures to include Space Elevators will enable a robust movement off-planet.

During the discussions we reached across the strengths of rocket launches along with their difficulties. We recognize there are three principal strengths: 1) rockets are successful today and great strides are forecast for the future, 2) reaching any orbit can be achieved, and, 3) rapid movement through radiation belts for people enables flights to the Moon and Mars. The strengths of a permanent infrastructure with daily, routine, environmentally friendly and inexpensive attributes come with Space Elevators. These strengths will be compared to the difficulties of executing a Space Elevator developmental program. Space Elevators will not be ready for initial human migration off-planet. However, once colonies are established on the Moon and Mars using rockets, Space Elevators will enable robust growth by moving massive amounts of cargo, daily, inexpensively, environmentally friendly, and routinely. The operational collaboration between the Galactic Harbour's Space Elevators and a variety of rocket launch competencies is essential. Together they will compose the space transportation architecture for this century. This future architecture enables movement of massive amounts of material for our journeys throughout the Solar System as well as ensuring movement of people rapidly through radiation belts. One obvious change in planning with this complementary architecture of both rocket portals and Space Elevator infrastructures is you don't need to depend heavily on the local resources at your destination to survive. With Galactic Harbours providing massive cargo movement, the restrictions of rocket deliveries no longer applies and the traveler can depend on support from Earth's facilities.

> The restriction of rockets, only delivering 2% of launch pad mass to interplanetary destination, turns into a strength of Galactic Harbour Earth Ports as 100% of lift-off cargo is released towards its destination. This relates to 100% of liftoff mass, 70% cargo and 30% reusable climber.

2.3 Galactic Harbour and Vision of the Future:

The Space Elevator story is still being written. The Apex is where the Galactic Harbour meets the shoreline of outer space and where the "Transportation Story of the 21st Century" meets the "Final Frontier."

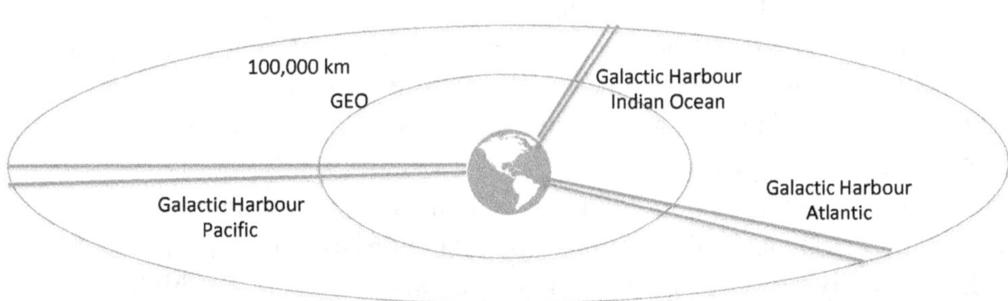

New Space Elevator Vision
Space Elevators are the Green Road to Space - they enable
humanity's most important missions by moving massive tonnage to
GEO and beyond. (safely, routinely, inexpensively, daily,
and they are environmentally neutral.)

2.4 Space Elevator Strengths:

ISEC's last report, "Space Elevators are the Transportation Story of the 21st Century,"
[Swan, 2020a] showed the magnitude of the change from rockets to a combined
architecture of rockets and Space Elevators moving mass for interplanetary missions. The
report showed the growth of Space Elevator transportation capabilities resulting in a mature
level of 170,000 tonnes of cargo per year easily going to GEO, Moon and Mars. This
inherent capability results from its "permanent" characteristics leading to a bridge into
outer space. Rockets are good for near-Earth and crewed missions because they are fast;
but, for interplanetary missions, Space Elevators win hands down. Of course, most
missions released from the Space Elevator will need rocket engines to slow down and
rendezvous with destinations. The new concept of Galactic Harbour Architectures has
unique characteristics that will "enable" interplanetary missions as complementary
infrastructure to rockets. These include:

- Routine Massive Lifts: During early operations, each Space Elevator Climber will
  carry 14 metric tonnes of payload to GEO and beyond with departures every day, or
  84 metric tonnes per day (14 x 2 SE x 3 GH) around the globe. This will happen
  365 days a year, or 30,660 metric tonnes per year to GEO and beyond. As the
  maturity is reached in massive liftoff Space Elevators, the number moves up to just
  less than 170,000 metric tonnes per year to GEO and beyond.
- Routine Daily Lifts: As the Space Elevator is designed to lift cargo daily, releases
  towards interplanetary missions will be standard and routine with no 26 month wait
  for periodic Mars alignment.
- Fast Transits to Mars Available: With the daily release of payloads towards Mars
  (and other interplanetary destinations) release from the Apex Anchor imparts
  tremendous velocity with very little drag from Earth's gravity. As a result, a
  periodic fast transit to Mars lowers the minimum time to 61 days. and could
  ultimately be as low as 40 days. [Peet, 2021]
- Reduction in Environmental Impact: As the tether climber ascends the Space
  Elevator, it receives energy from the sun and does not pollute along the way. The
  reduction of rocket launches to only the critical ones lowers their atmospheric
  impacts, hazardous particulate and greenhouse gas production, low Earth orbit
  debris, and pollution around launch sites.
- Promise to Planetary Scientists: Planetary scientific instruments, and their
  support equipment, can be assembled at the Apex Anchor with no restrictions of
  mass. In addition, daily releases can be achieved towards all planets at high
  velocity. [Peet, 2021]

In addition to last year's study, this study report expands the missions of Space Elevators to include being the Green Road to Space. It concludes that the construction and operations of Space Elevators will be carbon negative which essentially makes Space Elevators part of the "Big Green Machines" key to our planet's health. The two factors, Space Elevators are carbon negative and they enable Space Solar Power Satellites to replace hundreds of coal plants, embody that concept.

2.5 Summary of Customer Demand:

Development of new megaprojects are usually initiated because there is a tremendous customer need. As such, the authors have researched the programs that cannot succeed without the capability of Space Elevators. During a recent study [Swan 2020a], three destinations were chosen as reference missions: Space Solar Power, Mars Colony and Moon Village. The tremendous demands in terms of metric tonnes to support these customers makes it obvious that rockets alone will limit these missions. With cooperative activities tying rocket portals to Galactic Harbour infrastructures, these reference missions are possible within the desired time frames. To place this whole study in perspective, the comparison of "demand pull" for these three reference missions is identified as:

GEO Base -	Space Solar Power -	5,000,000 metric tonnes
Moon Base -	Lunar Village -	500,000 tonnes
Mars Base -	SpaceX Colony -	1,000,000 metric tonnes

In addition to those three significant programs for the near future, the International Academy of Astronautics produced a four-year study that identified more customer needs for movement of cargo to GEO and beyond. These are shown in Table 2.1 and combine with the other three Demand Pulls ensuring plenty of customer requirements for the next thirty years. If one were to look at the demand pull for GEO, the Moon and Mars, the future of cis-lunar growth depends upon the movement of massive tonnage - a strength of Space Elevators.

Table 2.1, Delivery Demand by Year in Metric Tonnes [Swan, 2014]

Demand in Metric Tons	2031	2035	2040	2045
Space Solar Power	40,000	70,000	100,000	130,000
Nuclear Materials Disposal	12,000	18,000	24,000	30,000
Asteroid Mining	1,000	2,000	3,000	5,000
Interplanetary Flights	100	200	300	350
Innovative Missions to GEO	347	365	389	400
Colonization of Solar System	50	200	1,000	5,000
Marketing & Advertising	15	30	50	100
Sun Shades at L-1	5,000	10,000	5,000	3,000
Current GEO satellites + LEOs	347	365	389	400
Total Metric Tons per Year	58,859	101,160	134,128	174,250

2.6 Space Elevator Throughput:

After discussing carrying capacity and operational dates of future Space Elevators in the last ISEC study [Swan, 2020a], projections of capability growth within the global transportation infrastructure is shown by the next figure.

Figure 2.3, Throughput for Future Space Elevator Architecture [Swan, 2020]

This development from a single Initial Operational Capability Space Elevator to three Galactic Harbours with the full capacity estimated to handle huge amounts of cargo illustrates the remarkable revolution in lift-off capability. The increase in this capacity over time is shown in Figure 2.3.

2.7 Efficiency of Natural Resources Usage:

When designing large rocket systems, the

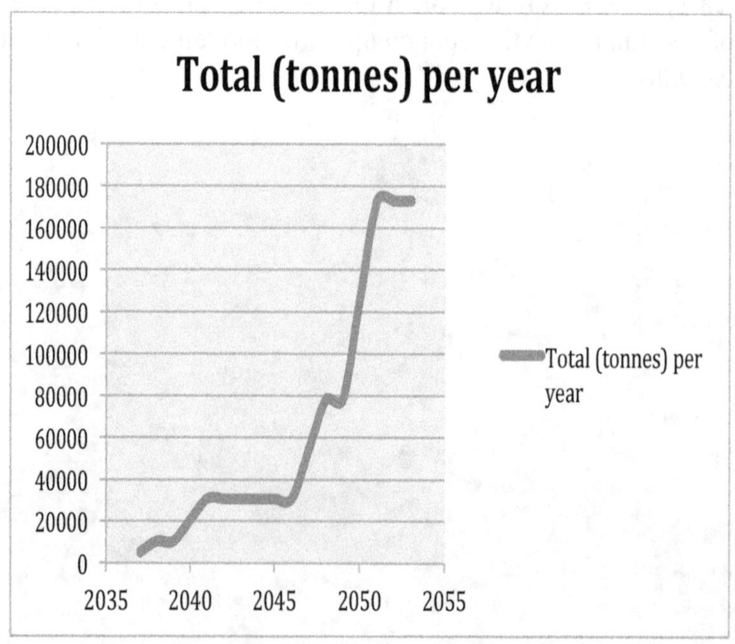

assumption is that the process will burn fuel to provide velocity and enable escape from the gravity well, as needed. This consumption of natural resources has been justified as a strategic need because of the remarkable scientific discoveries, orbital missions, and establishment of our permanent presence off planet. In the past, and for the near future, it is the only approach to fulfill our visions for orbital activities; however, in the future, the projection of rocket launches grows greatly when it is the only method of gaining height and speed. As a result, the consumption and burning, of natural resources will accelerate, resulting in much greater pollution. Can the efficiency of rockets still be accepted when the basic question is:

> How does a single space transportation infrastructure hope (rockets only) to aggressively move people and cargo off planet when they only deliver one half of one percent of the initial mass on the pad to their destination on the lunar surface? [Apollo, 1969]

2.7.1 The Rocket Equation Conundrum: The issue with rockets is that they are very poor, at delivering a high percentage of the original mass to destination. This major flaw of the rocket approach is the consumption of its initial mass at the pad to increase to the velocity required for orbital flight (approximately 9.4 km/sec of velocity required to just reach LEO). This consumption of pad mass is a huge portion of the total vehicle weight and decreases the payload capability for each launch. Essentially, to reach LEO, the rocket equation consumes 96% of pad mass (fuel burned and structure used). The remaining 4% is the payload mission equipment - everything else is released earlier for reuse, left in lower orbits as debris, or burned up to gain velocity. Reusability does not change the rocket equation delivery mass; it does make it more efficient and less costly. The reality is that 17,000 miles an hour to stay in LEO is demanding. Then, to gain velocity to go to the Moon, GEO, or Mars, the rocket equation demands consumption of even more fuel, structure, electronics and equipment along the way. The final velocity to GEO (or trans lunar or Martian injection) is hard to reach and those other parts of the rocket that do not contribute to the next stage of the mission are "thrown-away" while consuming fuel. Only two percent is sent towards high orbits such as GEO or lunar transfer. The real catastrophic number illustrating this point was the Apollo equipment that landed on the Moon (with Astronauts) represented less than half a percent of the mass on the launch pad at Cape Canaveral (see next table). This has not been improved upon with plans for future rockets burning fuel such as Starship and New Glenn.

Table 2.2, Destination Mass Delivery Percentages [Apollo, 1969]

Apollo Mission	Mass on Pad (tonnes)	Mass to Lunar Surface (tonnes)	Percentage of Launch Pass Mass %	Mass to Ocean Recovery (tonnes)	Percentage of Launch Pass Mass %
Saturn V	3,233	16.4	0.5	5.6	0.17

Consuming fuel, structure and equipment to gain velocity is a brutal approach - of course, it is the only approach today; but, it is still brutal. In addition, there are no "cost" or "reusability" factors in the rocket equation. You can do it more efficiently, but you cannot beat the 120 year old equation. These restrictions deliver miniscule percentages to destinations. They are only acceptable as there is no alternative. When Space Elevators are operational, the ability to deliver 100% of the liftoff mass to its destination on the tether will be standard. Looking at the SSP program puts this in perspective - Chapter 3 will expand on this whole topic. (also, see Appendix B on Avoiding the Rocket Equation)

2.7.2 Space Elevator Carrying Capacity: When Space Elevators arrive, they can begin to carry the burden of cargo movement and the consumption of fossil fuels can be reduced significantly. Space Elevators, by their design, climb to gain altitude and velocity using a reusable source of energy, the Sun. As such, transferring that energy into missions is a remarkable change. In addition, the efficiency of mass delivery to mission orbit is drastically improved when the move is made by Space Elevators. The following five estimates are used in this report to assess the choice of transportation alternatives inside a Dual Space Access Architecture. Space Elevators will deliver 70% of the mass at ocean surface and reuse the other 30%. The rocket equation drives the historic and future rocket numbers to draconian low mass delivery.

Table 2.3, Destination Delivery Percentages

Destination	Numbers
Space Elevator (100% to Tether destination - GEO or Apex Anchor 70% cargo and 30% tether climber to be reused)	70%
Rockets to LEO (reusable brings back first stages)	4%
Rockets to GEO & Trans-lunar trajectory	2%
Rockets to surface of Moon	0.5 %
Rockets to Mars (average 3 missions to Mars in 2020)	0.3%

Table 2.4 shows the delivery of cargo mass (mission payload) to mission orbit from a series of rockets (the data is a mixture of available data). The table validates the estimate used in our discussion; however, no matter the numbers used, the rocket equation dominates.

Table 2.4, Estimates of Delivery to Destination

Mission	Launch Vehicle	Total Mass at Pad (kg)	Mass at LEO Orbit	% to LEO Orbit	Mass at GTO Orbit	% to GTO Orbit	Comment
Spacecraft	Starship	5,000,000	100,000*	2.0	21000**	0.4	*Needs refueling to leave LEO, *for GEO no refuel
	NEW Glenn	1,323,529	45,000	3.4	13000	0.1	
Apollo	Saturn V	3,233,256	140000	4.3	41000	1.3	Tli vs. GEO
	Saturn V	3,233,256		0.5			To lunar surface
	Saturn V	3,233,256		0.2			Returned to Earth's ocean
	CZ-5-522	630,000	20,000	3.2	11000	1.7	
	Atlas V	590,000	18,500	3.1	8700	1.5	
Spacecraft	Ariane 5	737,000	20,000	2.7	10000	1.4	
	Soyuz	310,000	7,000	2.3			
	Soyus 2-1b Fregat	308,000	8,500	2.8	3000	1.0	
	Starship	4,000,000	100000	2.5	21000	0.5	Starship to GEO, no refueling
	Falcon Heavy	1,420,788	63800	4.5	26700	1.9	
Major Rockets	Averages			3.2		1.5	
Hope to Mars 2020	HIIA	350,000			1,350	0.4	fuel optimum
Mars 2020	Atlas V-541	531,000			1,025	0.2	fuel optimum
Voyager 1	Titan	632,970			1,820	0.3	to Jupiter then out of solar system

2.7.3 Space Solar Power (SSP) Example: With the demand for a SPS program to reduce environmentally damaging coal plant production, the Space Elevator is the only method that can enable a timely completion. Dr. Mankins stated that the project needs 5,000,000 tonnes of spacecraft moved to geosynchronous orbit to achieve the desired effect of supplying 12% of the global population by 2060 [Mankins, 2012]. Using Space Elevator transportation infrastructure, with daily, routine, safe, and inexpensive capabilities, the delivery will only take 28 years during development, versus the hundreds of years with conventional rocket delivery. This ability to raise mass with electricity avoids the catastrophic rocket equation and enables Earth-friendly liftoffs daily from multiple Space Elevators around the equator. If one were to look at the number of launches to place equivalent payloads to GEO, one would also be looking at significant impacts to the environment. The SpaceX Starship can move 100 tonnes to LEO, and advertises 21 tonnes to GEO without refueling.   To achieve customer demand for SSP alone, it would take the Starship 238,095 launches @ 3 per day = 79,365 days /365 or total of 217 years according to numbers derived from the SpaceX website estimates

2.8 Conclusions:

The question in the space arena should not be how do we build bigger and better rockets to support these customer demands, because the massive movement of cargo can never become efficient for rockets. Instead one must move to the Dual Space Access Architecture concept with Space Elevators moving massive tonnage while the Galactic Harbour encourages and develops space enterprises along their vertical "train tracks." Rockets have their place when delivering payloads to LEO. It is basically economical with minimal impact to the Earth's environment. However, when venturing beyond LEO to MEO, GEO and other planets it simply is not feasible to "build a bigger" rocket that has to make tens of thousands of launches to deliver the required payloads. Utilizing the capabilities of Space Elevators, coupled with rockets to create a "Dual Space Access Architecture" is the most efficient, cost effective way to deliver payloads outside Earth's neighborhood.

Chapter 3 – Enabling Space Solar Power

3.0 Beneficial Impact:

> Potential Beneficial Impact of Space Elevator: Enables Elimination
> of Coal plants with low-cost constant electricity from space.

3.1 Introduction:

Climate change is happening, and its effect on human habitats is projected to be substantial in the coming decades [Xu, 2020]. Contributing to this are the ever-increasing energy demands of the world's economies, which from current trends are projected to increase 40% by 2050, with a 70% increase in electricity demand via local power grids [USEIA, 2019]. Both climate change and the increasing demand for energy are problems that need to be solved simultaneously to ensure the future of a habitable, productive planet. Space-based solar power (SSP) presents such a solution.

Since the idea was first proposed by Peter Glaser in 1968, SSP has seen interest and support wax and wane over the decades [Glaser, 1968]. The basic architecture of an SSP system has remained roughly constant in this time, however, and consists of three primary components:

- an array of photovoltaics (perhaps coupled with mirrors) in orbit to collect solar radiation,
- a microwave or laser transmitter module that converts the collected energy into a beam directed to Earth's surface,
- a ground station with a receiver array that converts the received microwave or laser transmission into electrical power distributed via the local power grid.

The major advantages of such a system include continuous, high power output compared to terrestrial solar power deployments. If deployed to geostationary (GEO) orbit, such a system would not be subject to day/night cycles. Local weather or cloud cover of the ground station could be of minimal impact, at least for the case of microwaves as the transmission medium.

Full replacement of the world's current average electricity needs (as of 2014) with SSP would require over 2,400 GW of continuously delivered power (Figure 3.1). Even a fraction of this provided via SSP would be of enormous impact: increased electricity production from renewables in lieu of fossil fuels for baseload power is a key component of the global strategy to combat climate change [Pachauri, 2015]. Recently, Japan and China have committed to carbon-neutrality by 2050 and 2060, respectively [Denyer, 2020],

[China, 2020]. Other major fossil fuel consumers, such as South Korea, the United States, and the European Union, have declared similar intentions [Biden, 2020], [Cha, 2020]. [EU, 2019]. SSP, with its high output and availability compared to terrestrial solar power, can enable such ambitions.

However, despite economic arguments in favor of the idea [Mankins, 2013], one of the major obstacles to its implementation is the movement of a large mass and volume of hardware to GEO. Recent architectures require an estimated 3,500 to 25,000 tonnes of hardware deployed for a single system instance [Yang, 2016], [Mankins, 2013]. By comparison, less than 23,000 tonnes of payload hardware have been deployed to date by humans into orbit and beyond. Rocket-based Earth-to-orbit (ETO) delivery is unlikely to delives systems of such scale in any reasonable time.

A space elevator (SE) is the SSP deployment solution we propose in this study chapter. Originally conceived in 1895 by Konstantin Tsiolkovsky as a tower reaching to GEO [Pearson, 1997], this concept has evolved in technical maturity over the 20th and 21st centuries [Artsutanov, 1960], [Pearson. 1975], [Edwards, 2003], [Swan, 2013], [Swan, 2019]. Modern designs of such a system feature a thin ribbon under tension connecting an equatorial base station on Earth to an apex anchor beyond GEO at least 100,000 km in altitude. A process for the manufacture of the ribbon, requiring an unprecedented

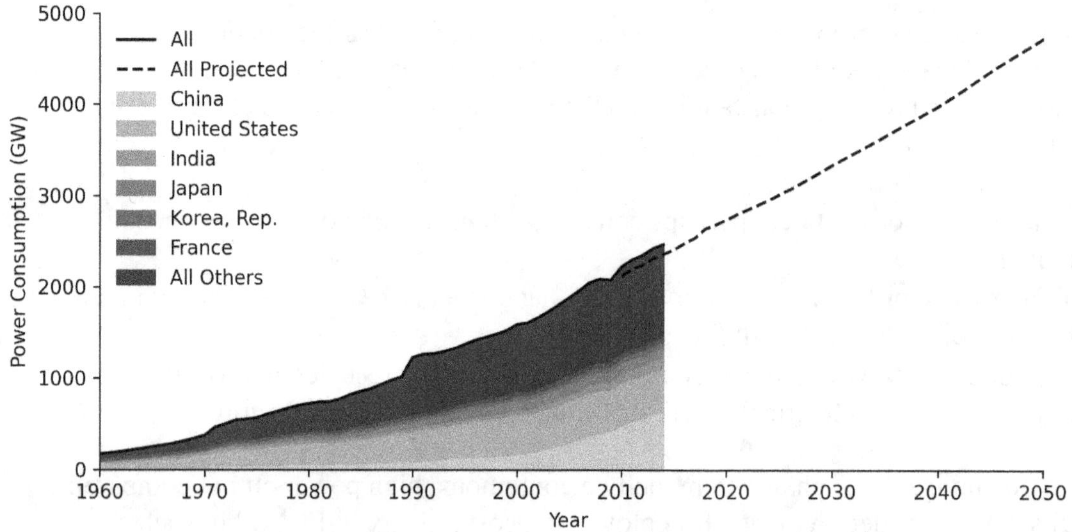

Figure 3.1: Global electrical power consumption measured in GW, time-averaged over usage within each year. Gradations in color give a qualitative impression of the distribution of usage over the 217 countries and territories included in the usage data. In 2014 the most recent recorded year available, this global time-averaged usage measured by its highest value ever, at 2.467 GW. Data derived using population and electricity consumption per capita from the World Bank Data Portal, [Swan, 2013], [WBDP,2020]. The superimposed "projected" line gives the modeled projection from 2010 to 2050 dev eloped by the U.S. Energy Information Administration. [USEIA, 2019]. The projected electrical power consumption at 2050 is estimated to be 4,730 GW.

tensile strength to mass ratio, is in development. However, with a near- constant projected transport throughput of 79 tonnes per day at full operating capacity for a single system instance [Swan, 2013], a space elevator is the transport system best suited for ETO deployment of mega-systems such as SSP,

We show in this study that it may be possible to deploy the first SSP system using rockets in the coming decades. If SSP is to become a major part of the renewable energy portfolio of the world's nations, however, the throughput of Space Elevators is necessary to deploy it at scale. Such throughput can deliver up to 1/8 of global electricity demand by 2070, and up to 1/4 around 2100, if resources are devoted to Space Elevator development now. The content of this study chapter asserts and supports two key points:

- Space-based solar power (SSP) is a necessary capability for the globe and can supply over 12% of electrical power demand by 2070.
- Space Elevators can deliver the massive hardware segments required for SSP assembly at geostationary orbit during their own development program.

3.2 Electricity requirements for existing cities and countries:

To discuss the feasibility of deploying SSP systems to provide usable power to terrestrial targets, we must establish the following:

- Typical power requirements for existing cities and countries, today.
- Power delivery and in-orbit mass of deployed SSP systems supplying that power.

We will focus on a selection of countries with high power requirements that could utilize SSP. These are given in Table 3.1. We have chosen these countries based on established interest in SSP by government and academic efforts. To get a sense of power requirements at a more local scale, we have also selected the five largest cities in the United States. These are given in Table 3.2

Table 3.1, Power Consumption by Country

	Population (M)	Time-Average Power Consumption (GW)
China	1,364	611.2
United States	318	471.9
India	1,296	118.9
Japan	127	113.5
South Korea	51	60.7
France	66	52.5

Average power requirements of a selection of countries that could utilize SSP, 2014. Calculated from publicly available population, electricity usage per capita data curated by the World Bank [WBDP 2014].

Table 3.2: Average power requirements of a selection of cities
that could utilize SSP, 2016. [NREL, 2016]

	Population (M)	Time-Average Power Consumption (GW)				Electricity Cost by Sector ($/kWh)		
		Total	Residential	Commercial	Industrial	Residential	Commercial	Industrial
New York	8.46	4.847	1.697	2.969	0.181	0.249	0.165	0.060
Houston	2.24	3.612	1.203	1.157	1.252	0.111	0.084	0.053
Los Angeles	3.92	2.697	1.032	1.372	0.293	0.154	0.144	0.119
Chicago	2.71	2.154	0.823	0.701	0.630	0.125	0.087	0.065
Phoenix	1.55	1.826	0.841	0.642	0.343	0.124	0.104	0.061

[Numerical figures obtained from the US Department of Energy LEAP and SLED data API maintained by the National Renewable Energy Laboratory.]

The discussion that follows is focused on meeting the electricity needs of these cities and countries today. This is a conscious choice, allowing for concrete discussion of the hardware required to meet real, historical demand. It is clear, however, that any discussion of SSP must be focused on meeting these needs in future decades. From Figure 3.1, it is clear that global electricity demand is expected to roughly double by 2050. However, on the scale of individual cities and countries, it is much harder to make detailed projections of such demand on these timescales. We will return to the question of meeting the global electrical needs of the future, through the year 2100, in Section 3.6.

3.3 SSP designs available to meet terrestrial demand:

To address the needs of these cities and countries, we have chosen to focus on a variety of proposed solar power satellite (SPS) system designs for SSP. These designs span 40 years of SSP research, allowing us to evaluate key figures of merit against our feasibility questions in this study chapter.

Table 3.3: Summarized specifications for a selection of SPS designs.

	Year Proposed	Orbit	Power (GW)	Transmit Frequency (GHz)	Mass (tonnes)	Ground Receiver Area (km^2)
Reference System	1979	GEO	5	2.45	30,000-50,000	—
SunTower	1995	LEO	0.1-0.4	5.8	2,000-7,000	—
Tethered SPS	2001	GEO	0.75	5.8	3,800	12.6
SPS ALPHA	2012	GEO	2	2.45	25,200	78.5
SSPS OMEGA	2016	GEO	2	5.8	23,000	78.5
SPS ALPHA Mk-II	2017	GEO	2	2.45	10,000	28.3
CASSIOPeiA	2017	GEO	0.43	5.8	400-900	7.84

[These proposed systems and their orbit, power, and mass specifications are used as the basis for discussion of ETO transport needs. We include the NASA Reference System [Hanley, 1978], Sun Tower [Mankins, 2002] , Tethered SPS [Sasaki, 2007] , SPS-ALPHA [Mankins, 2012] , SSPS-OMEGA [Yang, 2016] , SPS-ALPHA Mk-II [Mankins, 2017] , and CASSIOPeiA [Cash, 2017]. Table expanded from and inspired by Table 1 by Yang et al. [Yang, 2016].]

From these data, it is straightforward to assess the SSP requirements for our selected cities and countries. In all cases, we have chosen to represent power requirements in GW (gigawatts). Values given for each city/country are time-averages over measured electrical power consumption in the stated year, often given in either kWh (kilowatt-hours) or Btu (British thermal unit) in the source data. It should be noted that these values therefore do not represent peak power requirements.

The number of SPS systems of each design given in Table 3.4 needed to provide 1/8 (12%) of the time-average power required by each of the countries given in Table 3.2 is shown in Figure 3.4. For countries with lower power requirements such as India, Japan, South Korea, and France, this fraction of power needs could be met by a relatively small number (<10) of SPS-ALPHA/OMEGA systems. With the exception of India, these countries also have relatively small geographic areas available for terrestrial solar systems of comparable power output, offering an enticing alternative (see Appendix F).

For countries with the highest power requirements (China, United States), 30 to 40 such systems would be required. These two countries also have large geographic areas that could support substantial terrestrial solar deployment. However, it remains that the continuous power delivery of SSP systems without the need for battery storage may still make them a compelling alternative for these countries.

Complementing SPS-ALPHA/OMEGA systems with many smaller systems, such as CASSIOPeiA satellites, could offer flexibility for power delivery to different parts of the country based on seasonal requirements. For example, power from a CASSIOPeiA constellation could be diverted from warmer to colder regions of the U.S. in the winter months to provide power for heating, then diverted back in the summer months to hotter regions for cooling. Such changes in the terrestrial target for smaller SSP systems could prove less costly and more flexible than the existing approach to energy distribution in the U.S. by way of coal shipment and natural gas pipelines.

	SunTower	CASSIOPeiA	Tethered SPS	SPS-ALPHA	OMEGA	SPS-ALPHA Mk-II	Reference System
China	305.6	177.7	101.9	38.2	38.2	38.2	15.3
United States	235.9	137.2	78.6	29.5	29.5	29.5	11.8
India	59.5	34.6	19.8	7.4	7.4	7.4	3.0
Japan	56.8	33.0	18.9	7.1	7.1	7.1	2.8
South Korea	30.4	17.6	10.1	3.8	3.8	3.8	1.5
France	26.2	15.3	8.8	3.3	3.3	3.3	1.3

Figure 3.2: Number of SPS systems for each design to provide 1/8 (12.5%) of the power requirements for each examined country. Countries are ordered bottom-to-top by increasing power consumption. SPS designs are ordered left-to-right by increasing power delivery.

To break down requirements for the U.S. more locally, we have performed a similar analysis for its five largest population centers. The number of SPS systems for each design given in Table 3.3 to provide the full (100%) time-average power requirements for these cities as given in Table 3.2 is shown in Figure 3.3. Compared to the large numbers of SPS systems required to provide 1/8 of the power for the entire country, the number of systems needed for each of these cities is much less daunting. The largest consumer, New York City, would require a trio of SPS-ALPHA/OMEGA systems to cover its needs, or equivalently about a dozen smaller CASSIOPeiA systems. The other cities featured in this list could provide a substantial fraction of their power requirements on a single SPS-ALPHA/OMEGA system, or all their requirements on a pair of systems.

Meeting the full power requirements (albeit time-averaged, and not factoring in peak loads) is the extreme case for these examples. More likely, for these cities SSP would be complemented by wind, hydroelectric, terrestrial solar, and nuclear power in providing electricity at the levels needed to support varying load over the course of a year. As such, SPS systems supporting these cities would be able to supply consistent power to consumers in surrounding counties as well, supplemented by a smaller complement of natural gas plants.

More numerous, smaller SPS systems, such as CASSIOPeiA systems, could be utilized to provide significant fractional support for each of these cities. As noted previously, more numerous, smaller systems can afford greater flexibility in power delivery, transmitting as demanded geographically.

Phoenix requires substantially more electricity during its hot summers than it does during its mild winters. Its large residential footprint adds to this seasonality in usage. For the summer months, an SPS system (or many smaller SPS systems) could provide supplemental electricity to the city, reducing or eliminating the need for the same power delivered via coal-fired plants. When winter arrives, the same SPS system(s) can be redirected to other cities further north or at higher altitudes (for Phoenix, perhaps Flagstaff), providing the supplemental power they need for heating.

	SunTower	CASSIOPeiA	Tethered SPS	SPS-ALPHA	OMEGA	SPS-ALPHA Mk-II	Reference System
New York	19.4	11.3	6.5	2.4	2.4	2.4	1.0
Houston	14.4	8.4	4.8	1.8	1.8	1.8	0.7
Los Angeles	10.8	6.3	3.6	1.3	1.3	1.3	0.5
Chicago	8.6	5.0	2.9	1.1	1.1	1.1	0.4
Phoenix	7.3	4.2	2.4	0.9	0.9	0.9	0.4

Figure 3.3: Number of SPS systems for each design to provide the full power requirements for each examined US city. Cities are ordered bottom-to-top by increasing power consumption. SPS designs are ordered left-to-right by increasing power delivery.

3.4 Deploying SSP by rocket:

Given the substantial amount of hardware required at GEO to provide countries and cities with space-based solar power, the highest barrier to deployment is Earth-to-orbit (ETO) transport. Rockets continue to be the only such transportation system available at this time. However, the cost of rocket based ETO is decreasing through reusability efforts of commercial launch companies, in particular SpaceX and Blue Origin. As the barrier lowers, does it lower far enough for SSP to become feasible?

The cost of deployment using rockets has been explored by other authors [Mankins 2013]. A question less considered is the number of rocket launches required to deploy systems on the scale of an SPS. This is important, as massive increases in the scale and frequency of rocket launches may have a measurable, negative impact on the environment, including the climate. It is also the case that even with reduced launch cost, the throughput of ETO hardware deployment may simply be insufficient to meet the needs of the cities and countries discussed in Section 3.2 in reasonable time.

Table 3.4: Cargo and launch masses for a selection of reusable rocket delivery systems bound for GTO. Figures drawn from user guides published by SpaceX and Blue Origin.

	Cargo Mass (tonnes)	Launch Mass (tonnes)
SpaceX Falcon 9	8.3	549.0
SpaceX Falcon Heavy	26.7	1420.8
SpaceX Starship	21.0	5000.0
Blue Origin New Glenn	13.6	–

To consider the question of throughput, let us assume a heavy launch vehicle capable of delivering 20 tonnes to GTO. This value is representative of current and upcoming heavy launch systems from industry players given in Table 3.4. To simplify our discussion further, let us focus here on the requirements for deployment of a single SPS-ALPHA Mk-II, the lightest of the 2 GW systems (and the most efficient in land usage, see Appendix F). From Figure 3.3, a single such system would support most of the power requirements for a large city such as Chicago or Los Angeles.

Deploying a single SPS-ALPHA Mk-II (10,000 tonnes) using a 20-tonne GTO launch vehicle would require at least 500 launches. As a point of comparison, there have only been about 100 successful launches globally each year for the past three years, and few of them heavy launch vehicles of this type. But the rocket industry is increasing its pace. SpaceX and Blue Origin in particular are building larger, reusable launch vehicles to support such payloads, with increasingly aggressive launch schedules as an intended outcome. Three launches per day is Elon Musk's stated goal for SpaceX's Starship in support of Mars colonization [Musk, 2020], amounting to over 1000 launches per year.

With such launch frequencies in play, it becomes possible to launch an SPS system on the scale of SPS-ALPHA Mk-II. However, it remains unclear how much capacity would be available to such projects, and when. An order of magnitude increase in launches over current global totals is required for the Mars colonization effort alone.

Although deploying a single instance of a 2 GW SPS system would finally be within the realm of possibility, satisfying a significant fraction of the world's power requirements with SSP would require far more. Nearly 100 such systems would be needed to deliver 1/8 (12.5%) of today's electricity requirements for all of the countries listed in Table 3.1; over 150 systems would be needed to do so globally (Figure 3.1). That ambitious milestone would itself require 75,000 heavy launches, or 75 years of Musk's already-ambitious Mars throughput.

And this is only to meet today's needs. In Section 3.6, we show that substantially more hardware will be needed to meet the same fraction of global demand in the coming decades. We also show that it is entirely possible to do so.

3.5 Deploying SSP by Space Elevator:

Rockets are incredibly useful, propelling our entry into orbit and beyond for the full measure of mankind's space activities to date. However, they are fundamentally constrained by the rocket equation, resulting in only around 2% of the total mass at launch available as payload for an orbit at GEO. This draconian constraint makes massive hardware movement from the gravity well of Earth to GEO incredibly resource-intensive with rockets.

An alternative deployment approach is the use of Space Elevators (SE). These systems are not constrained by the physics of the rocket equation, and in principle could deliver mass to orbit at the throughput required for large-scale SPS deployment. Table 3.5 gives the two such system types proposed by ISEC [Swan, 2020], [Swan, 2013], [Swan, 2019].

Table 3.5: Cargo and launch masses for a selection of
reusable rocket delivery systems bound for GTO.

	Cargo Mass (tonnes)	Lift Mass (tonnes)	# Climbers per Tether
ISEC IOC	14.0	20.0	7
ISEC FOC	79.0	100.0	7

Space elevator architectures proposed by ISEC, including "initial operating capacity" (IOC) and "full operating capacity" (FOC) designs. Lift mass gives the fully-loaded total mass of a climber with cargo. The mass columns correspond to those given for rockets in Table 3.4.

Once more choosing the 2 GW SPS-ALPHA Mk-II as our basis for discussion, we can quantify the rate of deployment of such a system given ISEC's designs. ISEC's current vision calls for a single IOC SE by 2037, followed immediately by a second in 2038 [Swan, 2020], [Swan, 2019], as shown in Table 3.6. By 2041, two more pairs of elevators are expected to come online, with a total capacity at six "initial operating capacity" (IOC) Space Elevators. The combined throughput of these six IOC Space Elevators would give up to 84 tonnes of hardware to GEO per day, at 14 tonnes per climber "lift" occurring each day on each elevator. If devoted to the deployment of a single SPS-ALPHA Mk-II, it would take 119 days to complete delivery. Such throughput would allow up to three (3) such systems deployable in a year, the full electrical power needs of New York City today (Figure 3.3).

Once more choosing the 2 GW SPS-ALPHA Mk-II as our basis for discussion, we can quantify the rate of deployment of such a system given ISEC's designs. ISEC's current vision calls for a single IOC SE by 2037, followed immediately by a second in 2038 [20,21], as shown in Table 6. By 2041, two more pairs of elevators are expected to come on-line, with a total capacity at six "initial operating capacity" (IOC) SEs. The combined mass deliverable to orbit per day across all elevators in operation is given as the throughput. These values aggregated over a full year are given as the annual max mass delivered, while the cumulative max mass delivered is an accumulation of these values with each increasing year. These values give an upper bound on the hardware delivery capabilities of the combined systems.

This is a remarkable statement, but this throughput is not a hard ceiling. By 2051, it is ISEC's vision that these six Space Elevators be upgraded to "full operating capacity" (FOC). With 79 tonnes per climber "lift" each day on each elevator, the combined throughput of these six FOC SEs approaches 474 tonnes per day, over 5x the rate for the six IOC elevators. This reduces the minimum deployment time of a single SPS-ALPHA Mk-II to a mere 21 days, or three weeks. If used continuously for a year, 17 such systems could be deployed. At this point, assembly of the deployed hardware may take longer than transport to GEO and become the new bottleneck for an operational SPS.

The impact of such transport capacity to GEO is difficult to overstate. Large hardware systems such as an SPS-ALPHA Mk-II no longer remain distant dreams by virtue of their mass, and they can be deployed in such numbers that their impact on energy and climate needs becomes substantial. A faster timetable for deployment manifests as greater impact for these systems, improving their chances of being pursued at all.

Reducing or removing the constraint of the mass of a system in GEO also allows for economies of scale in SPS design. Is it possible to design an SPS that delivers 5x the power (10 GW) of an SPS-ALPHA Mk-II, but with only a 3x increase in mass? What other aspects of the SPS design were seriously constrained by hardware mass, but can now be relaxed to great benefit for system performance? Such questions no longer put the goal even further out of reach and can instead be seriously considered.

Table 3.6: ISEC Space Elevator deployment schedule [Swan, 2019b], [Swan, 2020a]

Year	n_{IOC}	n_{FOC}	Throughput (tonnes/day)	Annual Max Mass Delivered (tonnes)	Cumulative Max Mass Delivered (tonnes)
2037	1	0	14	5110	5110
2038	2	0	28	10220	15330
2039	2	0	28	10220	25550
2040	4	0	56	20440	45990
2041	6	0	84	30660	76650
2042	6	0	84	30660	107310
2043	6	0	84	30660	137970
2044	6	0	84	30660	168630
2045	6	0	84	30660	199290
2046	6	0	84	30660	229950
2047	5	1	149	54385	284335
2048	4	2	214	78110	362445
2049	4	2	214	78110	440555
2050	2	4	344	125560	566115
2051	0	6	474	173010	739125

ISEC Space Elevator deployment schedule; nIOC and nFOC give the planned number of IOC and FOC elevators in operation in a given year. The combined mass deliverable to orbit per day across all elevators in operation is given as the throughput. These values aggregated over a full year are given as the annual max mas delivered, while the cumulative max mass delivered is an accumulation of these values with each increasing year. These values give an upper bound on the hardware delivery capabilities of the combined systems.

3.6 Meeting future global electrical demand with SSP:

We have shown that delivering SPS systems at scale can be achieved with the ISEC Space Elevator development program. But is the rate of delivery high enough to meet the problem of climate change in this century, producing renewable electricity at scales sufficient to meet the growing demand of the future?

To answer this question, we return to the model projection given in Figure 3.1, developed by the U.S. Energy Information Administration [USEIA 2019]. This projection gives an estimate of global electricity demand spanning the years 2010 to 2050. For the years with measured data (2010 through 2014), it is accurate to within a few percent of the known values. Over its whole domain of time it is also fairly linear.

We extrapolate this model forward to 2100 with a linear least-squares fit, dividing by 4, 8, and 16 to estimate 1/4, 1/8, and 1/16 of global demand, respectively (Figure 3.4/3.5). We are interested in electricity needs beyond 2030, by which time a rocket launch schedule such as that for SpaceX's Mars colonization program could be operating at full scale.

In Section 3.4, we argued that deploying SSP systems by rocket would be possible, although likely not at the scale required to meet global demand to a high degree. Assuming three launches per day of the 20-tonne-to-GTO launch vehicle we examined in Section 3.4, we superimpose the electrical power in GW that could be deployed over time by such a launch program were it fully devoted to deploying SPS-ALPHA MK-II systems (Figure 3.5). As we observed in Section 3.4, such a schedule could deliver about 2 SPS-ALPHA Mk-II systems per year, or up to 4.4 GW/year at 1,095 launches annually. This would give 310 GW of SSP-delivered power in orbit by 2100. However, this would amount to less than

4% of global electricity demand, which by 2100 will have swelled to 7,886 GW based on our extrapolation.

By meeting such a low fraction of the world's demand by 2100, SSP would scarcely contribute to the renewable energy portfolio required to mitigate climate change in this century. This low impact, even with an aggressive launch schedule, would serve to make SSP an unlikely candidate for any climate change mitigation program. SSP would be limited to serving the needs of countries or cities for which its unique features make it an appealing option over other renewable choices.

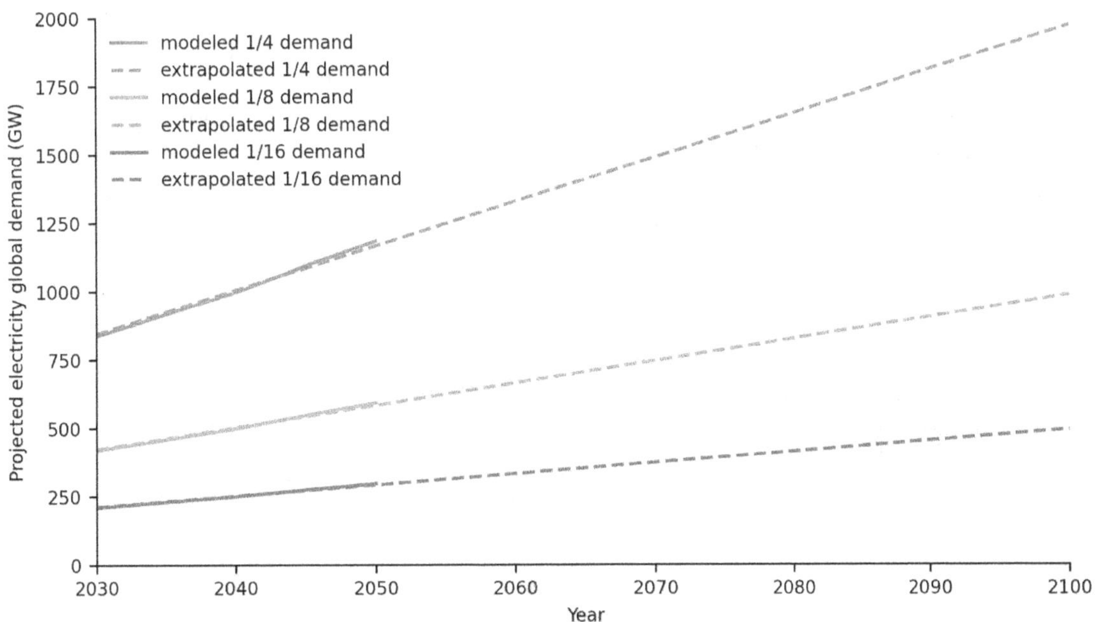

Figure 3.4: Modeled projection (solid) developed by the U.S. Energy Information Administration [USEIA 2019]. This is the same model projection displayed in Figure 3.1. Dashed lines give a linear least-squares fit applied to this model, extrapolated to 2100.

However, if we consider the throughput provided by the space elevator program proposed by ISEC (Table 3.6), a far more optimistic picture emerges. Assuming the full throughput of the program is devoted to SPS-ALPHA Mk-II deployment (unlikely, but the aggressive case), we observe that by 2056 space elevators will have delivered 1/16 (6.25%) of global electricity needs. At this point, the set of 6 "full operating capacity" (FOC) elevators would be delivering 17 SPS-ALPHA Mk-IIs per year to orbit. This amounts to over 34 GW/year, or nearly 8x the delivery rate of the 20 tonne launch vehicle program (at 1,000 per year).

By 2068, 3.6M tonnes of deployed hardware will increase the total electrical power capacity delivered by SSP to 1/8 (12.5%) of global demand, or 728 GW. Achieving this significant fraction would position SSP as a viable renewable energy source for the global climate change mitigation strategy, allowing SSP to join the portfolio of other renewable technologies to meaningful impact in this century.

By 2100, the total SSP power capacity required to deliver 1/8 (12.5%) of global demand will have grown to over 980 GW, requiring nearly 490 SPS-ALPHA Mk-II systems, or 4.9M tonnes at GEO. However, at the throughput of hardware delivery available from the six FOC space elevators, the number of such SSP systems will have grown to supply nearly 1/4 (25%) of the world's electrical needs. This milestone will involve 9.2M tonnes hardware, or over 920 SPS-ALPHA Mk-II systems, to achieve, but the impact is enormous. At this point, renewables should dominate the electrical power mix of sources; SSP can be dominant among them, delivering consistent solar power to cities and countries the world over.

This analysis assumed a constant frequency of rocket launches per year starting in 2030 and proceeding through 2100. Of course, if three launches a day of a 20 tonne launch vehicle is possible to support a single program in 2030, it is likely that far more launches are possible as the decades progress. So, our projection for rockets may be optimistic for 2030, but pessimistic in 2070 and beyond. However, to meet 1/8 of the world's electrical needs by 2100, over 3x this number–around 3,500 launches–would be needed each year starting in 2030. To match the throughput of the six FOC space elevators, over 8,600 launches a year would be required, an 8x increase over the assumed schedule.

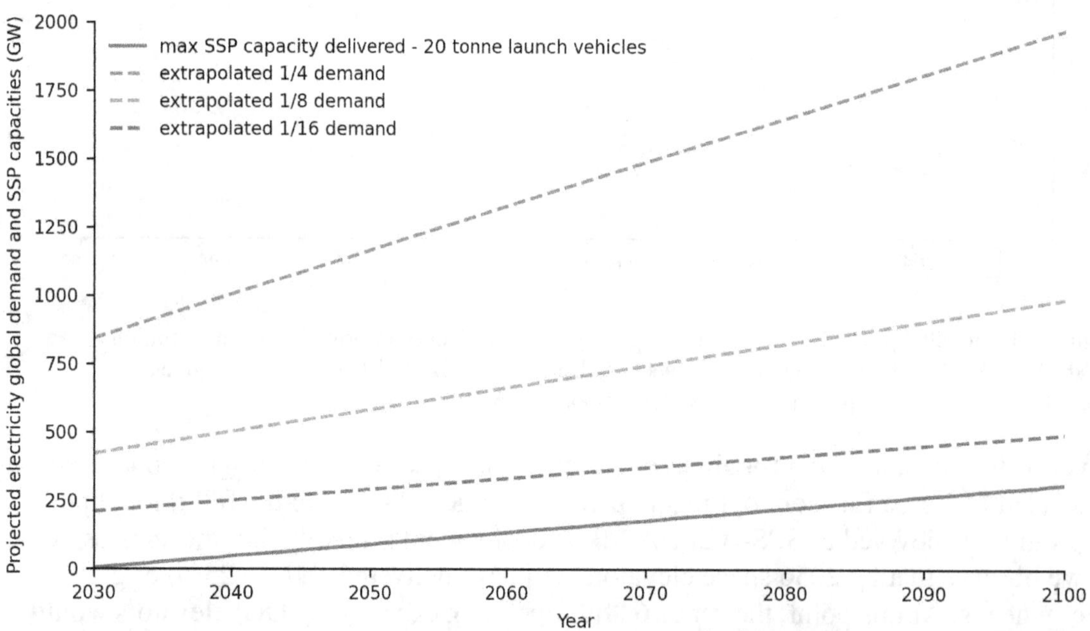

Figure 3.5: Modeled projection (solid) developed by the U.S. Energy Information Administration [Energy Analysis 2019]. This is the same model projection displayed in Figure 3.1. Dashed lines give a linear least-squares fit applied to this model, extrapolated to 2100.

In the same vein, in assuming the ISEC deployment schedule for space elevators, we have not considered the deployment of new elevators beyond 2051. Barring any fundamental constraints on elevator placement, it is unlikely that deployment of space elevators would end abruptly at six. New elevators, improving upon the designs of the existing ones, would surely come online over the decades to follow. The remarkable throughput examined here shows us what is at least possible when unconstrained by the rocket equation, and pursuit of a space elevator development program now opens up these possibilities to immense, global effect.

3.7 Complementary technologies – a Dual Space Access Architecture:

Transporting the massive amount of hardware to GTO required for the first SPS will likely be possible with the next generation of reusable heavy launch vehicles, operating on the frequent schedules intended by SpaceX and Blue Origin. Rocket technologies are vital to opening up new space activities, as they have been for humanity's entire history beyond the atmosphere. Perhaps most important, rockets are the only system capable of moving humans quickly to their destinations in orbit, with little time exposed within the hazardous Van Allen radiation belts.

To enable activities that are mass-intensive, but do not require movement of humans, space elevators function as a powerful complement to rockets. Once established with an initial system deployed by rocket, an SPS such as the SPS-ALPHA Mk-II could be deployed in massive numbers with high-throughput space elevators to service the needs of diverse cities and countries. They could even be deployed at such scale to become an integral part of the global climate mitigation strategy in the second half of the 21st century.

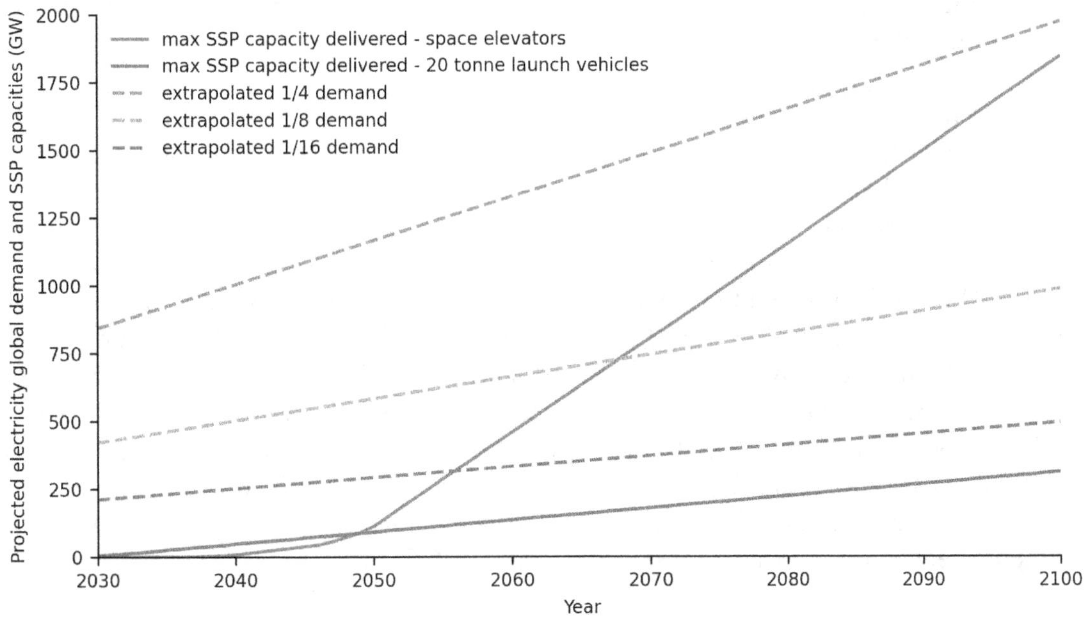

Figure 3.6: Cumulative SSP electrical power (GW) deployed by: (a) Space Elevator, using the full capacity of the ISEC development schedule given in Table 3.8 (solid violet); (b) a 20 tonne launch vehicle, launching three times per day, starting in 2030 (solid red). Dashed lines give a linear fit extrapolating to 2100 a model for global electricity demand. The original model was developed by the U.S. Energy Information Administration, spanning 2010 to 2050. [USEIA 2019]

The use of Space Elevators will move the massive freight required to support humanity's expansion into space. This will reserve rocket capacity for human-centric and exploratory movement, enabling rockets to continue increasing the presence of humanity in space. These complementary technologies form a dual space-access architecture that will secure our foothold beyond Earth's atmosphere.

3.8 Conclusions:

Space-based solar power (SSP) holds enticing promise as a flexible, renewable energy system for meeting the substantial electricity demands of the future while mitigating the negative impact on Earth's climate. However, SSP deployment via rocket alone remains constrained by the physics of the rocket equation, which fundamentally limits the rate at which massive hardware can be delivered to orbit. In this study chapter, we argued the necessity of using Space Elevators for deployment of SSP systems at scale to meet the growing demands for electrical power on Earth.

We have shown that for many cities and countries, SSP provides a promising approach to meeting the electricity needs at a reasonable scale. Fewer than ten SPS ALPHA Mk-II systems could supply 1/8 (12.5%) of the electrical power needs for each of India, Japan, South Korea, and France. For China and the United States, over thirty such systems would be needed to supply this fraction of electricity demand, but the advantages of lower land usage per GW and substantial, consistent supply of solar power across geographies make it an appealing alternative to other renewables.

We showed that the picture for cities is favorable to SSP, even for large ones: for New York, the largest city in the U.S. by electricity usage, only three SPS ALPHA Mk-II systems would be needed to meet 100% of its typical electricity needs. For each of the next four cities of greatest demand (Houston, Los Angeles, Chicago, Phoenix), less than two such systems are required. The ability of SPS to direct power to different locations in a region as needed could be utilized to provide power flexibly to these cities with demand seasonality, further reducing the need for fossil fuel consumption at power plants nearby. However, the problem of deployment has stood as a barrier to SSP since its conception by Peter Glaser in 1968. The massive hardware requirements of a single SSP system, such as the SPS ALPHA Mk-II, remain prohibitive. The aggressive launch schedules being pursued by industry players such as SpaceX and Blue Origin, however, are likely to make this possible in the coming decades, perhaps as early as 2030. We believe rockets will deliver the components needed for the first SSP system to orbit at GEO.

But if SSP is to be delivered at the scale needed to address the climate crisis, providing a large fraction of the world's growing electricity demand, delivery by rocket is likely to fall short. Massive movement of hardware at high, reliably constant throughput is required, and space elevators stand to meet this need. We have shown that the ISEC development schedule could enable this movement, delivering up to 17 SPS ALPHA Mk-II systems to orbit each year at full operational capacity. Global electricity demand could be met by SSP at substantial fractions this century, with 1/16 provided by 2056, 1/8 by 2068, and 1/4 around 2100 if an aggressive SSP program is pursued using space elevator resources. Any hope of achieving this requires initiation of concerted space elevator development now in order to achieve the first lifts of hardware by 2037. There will always be a place for rockets in mankind's pursuit of a presence in space, opening up new activities and habitats. But for massive movement of hardware, a daily, routine, high-throughput infrastructure is needed. Space elevators present such an infrastructure and are a linchpin for unlocking space for mega-systems such as SSP for economic and environmental impact at the global scale.

Chapter 4 - Environmental Benefits of the Space Elevator - Permanent Disposal of High-Level Nuclear Waste

4.0 Introduction

Humanity has recognized that there are some daunting problems associated with 20th Century technological advances. Developing nuclear weapons of mass destruction resulted from the global scale of warfare and industrial countries infinite demand for power presented problems that were not contemplated. High-Level Nuclear Waste from nuclear weapons and nuclear power plants was identified early as a problem; however, the problem grew and became pressing and expensive. A solution was obviously needed.

1. Nuclear reactors safely generate enormous amounts of inexpensive energy without adding greenhouse gases, but also create massive amounts of high-level radioactive waste.
2. There has been, and continues to be, significant nuclear waste resulting from weapons development, testing and disposal.

The Hanford DOE Site, (in Washington State) was one of the four main U.S.A. sites for the production of uranium and plutonium to be used in nuclear weapons. During the early 1940's at the Hanford site (but quite possibly everywhere in the U.S. at that time), the exact nature of the dangers the workers faced was not stated, but strict safety protocols were employed. Low-level radioactive waste was often in the form of radioactive dust in the air, or, in liquid-form discharged either underground or directly into the nearby Columbia River [Gerber, 1992]. The clean-up of the Hanford site officially began with the 1989 signing of the Hanford Tri-Party Agreement [Agreement, 1989]. By the end of 2019 [Newcomb, 2019], the clean-up may have already cost U.S. Taxpayers over 140 billion dollars, with estimates of another two to four times that amount still to be required. Similar situations, (to varying degrees), exist at the other U.S. Sites as well as in Russia, the United Kingdom, China, and other Nations that have developed nuclear weapons. One of the early proposals for 'permanently' disposing of this H-L-W was made by NASA in 1978 [Burns 1978]. Many options, (high Earth orbits, lunar orbits, solar orbits or even Solar System escape), were carefully investigated. However; the horrendous consequences of "sub-orbital launch failures" plus the draconian economics of the "rocket equation", [Siegel 2019], have led to the abandonment of such proposals. Fortunately, the Space Elevator will re-introduce these topics as being very "doable" and "Green." For high-level nuclear waste, the problem is simple to describe: Disposal, of hundreds of thousands of tonnes of highly radioactive waste from around the globe, must be accomplished in an environmentally friendly manner. In addition, it must be permanent, as much of the waste has extremely long half-lives.

Proposal in Two Parts:

- Part A: Recognize and Accept: Permanent disposal of High-Level Nuclear Waste is a monumental problem and currently has minimum success for safe and long-term storage.

- Part B: Recognize the Opportunity: Space Elevators could safely send the High-Level Nuclear Waste into orbits around the Sun that would never be near humans again.

4.1 General Information

This monumental environmental problem is international and results in radioactive waste sites dispersed around the globe. The problem must be addressed as there are three major current driving functions demanding action:

Significant Supply of Energy: 29% of low-carbon electric energy [iea, 2018] around the globe is produced by 441 nuclear generators today [WNA, 2020], with an estimate of a 60% increase by 2040[https://youtu.be/KqSmDJGgfTU]. This estimate does not consider an extensive development of space-based solar power.

- Leftover weapons: The global military competition resulted in over 64,000 nuclear warheads during the peak of the cold war in 1986 [peak, 1986] which was reduced to around 13,865 by 2019[].

- The permanent disposal of high-level nuclear waste (from both sources - weapons and power) has had problems; it is not proceeding as rapidly as hoped; and, still needs some resolution as to the best way to handle it. There are significant issues with disposal of high-level nuclear waste as the dangers are real and the engineering suggestions depend on so many diverse aspects.

We will discuss some of the problems, show the current method of disposal of high-level nuclear waste across the globe, and then show how the Space Elevator can contribute to a safer and more permanent solution.

4.2 Background

There are three parts to this section: definitions of the problem, description of the power generation waste problem and a description of the military warhead problem.

4.2.1 Nuclear Fission

 Nuclear fission processes generate nuclei that are unstable (radioactive fission products). The resulting radioactive decay, (alpha, beta or gamma), poses severe threats to our health and our environment when not isolated and chemically stabilized. This chapter will estimate the world-wide quantities of High-Level-Radioactive-Waste, (HLW) that has resulted from both the production of Nuclear Weapons, (between 1943 and the 'peak' of about 1985), and from the operation of commercial Nuclear Electric Generators, (1954 until the present, 2020, plus some tentative projections to the year 2050)[]. Note that we will use the abbreviations t for tonne, (1,000 kilograms), Kt (for 1,000 tonnes), and Mt (for mega-tonnes, 1 Million tonnes).

4.2.2 Nuclear Electric Generators: The beauty of nuclear generators is that they are able to generate about 560,000 times as much energy as the burning of crude oil [crude oil 2020], (or 200,000 times as much energy as the burning of pure Hydrogen gas), as well as during their actual operating-lifetime they emit zero Carbon Dioxide [near zero]. As well, nuclear generated electricity is among the least expensive methods of generating electrical energy [low cost]. Nuclear Reactors (App. E.2c) have, by far, (1) the lowest rate-of-deaths-from operations, and 2) 'footprint' - land use for both fuel-mining and electricity production - of any form of commercial electric energy production.   However, the left over nuclear material is still radioactive with the high-level waste being only 3% of the total volume, but demanding near-perfect isolation and monitoring.

In the future, Generation IV Reactors [Gen IV] will result in decreases in both the quantity of high-level waste and the length of time it must be sequestered (App E. 2d). However, these are just starting construction in 2020 so it will be several decades before they start to have the revolutionary impact that seems almost assured.

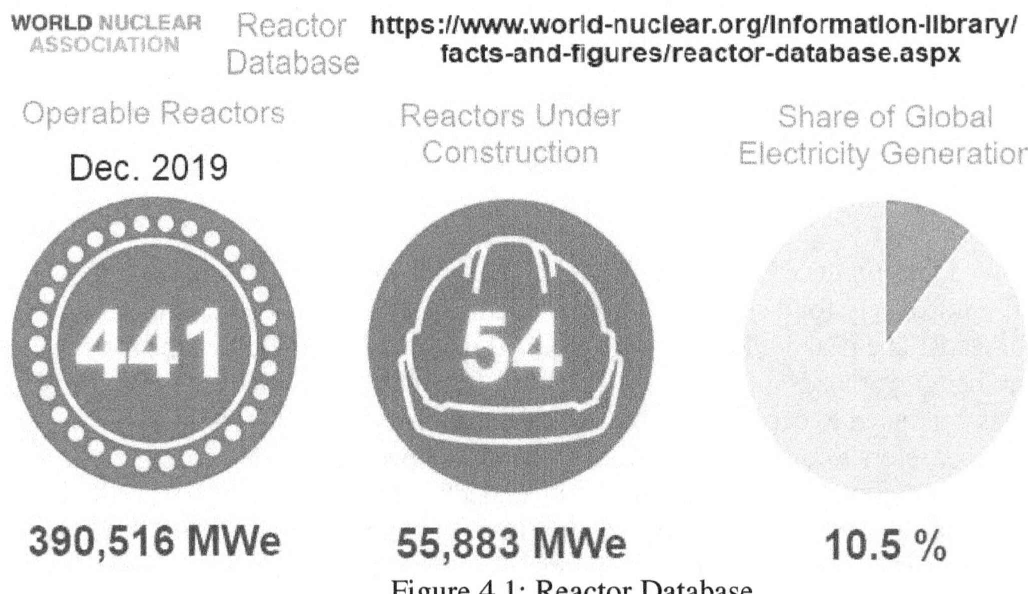

Figure 4.1: Reactor Database

One of the significant problems is that 'spent-nuclear-fuel-bundles', (SNF), that can no longer safely remain in the Reactor and be of significant economic benefit, are removed and stored under several meters of circulating water in an extremely large pool, constructed from reinforced, metal-lined concrete (App E 2e). The pool is usually adjacent to the Reactor Building. This is commonly known as 'wet' storage [wet]. After several years, (or even decades), the SNF is either sent to a reprocessing laboratory [reprocess], or it goes to a 'repackaging' facility [repackage], and then to 'dry' storage [Dry]. Reprocessing the SNF leads to newly usable fuel, (which needs to be enriched in Uranium before it can be used). The reprocessing also leads to H-L-W products of only about one-fifth the original volume [one fifth], but still dangerously radioactive for several hundreds of millennia1 . A summary table of the current situation, from 1954 until the end of 2018, is shown in the Table 4.1 below (Please see App. E 2e for the details plus estimates of increases, based on the planned expansion by several Nations for the periods 2019 through 2030 and 2031 through 2050).

---

1 vitrified_hlw.pdf; (pages 1 through 4) 22 PUB1822_web.pdf (2019) pp 3-4

Table 4.1, Summary of Expansions (Vitrified is for 'Calcined')

Type of Highly Radioactive Material	Storage Method	'World' Total Amount, (Kt)	Planned 'Future' for the Material
'Vitrified'*	Dry	36.5	Permanent Disposal
'Repackaged'	Dry	145	Permanent Disposal
'Spent-Fuel'	'Wet'	190	Unknown
'Vitrified'* (increase)	Dry	32.8	Permanent Disposal
'Repackaged' (increase)	Dry	90	Permanent Disposal
'Spent-Fuel' (increase)	'Wet'	50	Unknown
'Vitrified'* (increase)	Dry	43.8	Permanent Disposal
'Repackaged' (increase)	Dry	120	Permanent Disposal
'Spent-Fuel' (increase)	'Wet'	88.4	Unknown

4.2.3 Nuclear Weapons produce high-level Waste[]: The 'arms race' with Russia [Proliferation] immediately followed WWII and resulted in the making of an enormous number of warheads. The peak in the arms race resulted in more than 64,000 warheads by 1986. The number of warheads in 'service' in 2014 was less than one-sixth of the 1986 'peak', and it has continued to decrease to a 2019 'stockpile' of 'only' 3,800 [warheads 2018]. This number refers to 'actively deployed warheads'. However, also according to the 'Stockholm Institute' the total number of warheads had declined to 13,865 in 2019 [13,865 2019].   The welcome decrease since 1986 of over 50,000 in the number of warheads, along with the earlier insane rush from 1945 to actually detonate over 2,000 warheads to test their operational efficiency [testing nukes], has inevitably led to large amounts of H-L-W in the U.S., Russia, the U.K., France, China, and the other three known Nuclear Nations. Shockingly, from 1946 until it was formally outlawed in 1994, thirteen nations actually disposed of low-level and intermediate-level radioactive waste at several ocean locations [iaea dumping]. Only Russia has admitted to extensive ocean-dumping of H-L-W, but claims it was all done by the former U.S.S.R. [USSR]

One recently developed, but now widespread, method for chemically immobilizing H-L-W involves heating the waste and 'glass-forming materials' to an extremely high temperature until 'liquid glass' is formed. This is then poured into steel containers that are permanently sealed [vitrified]. The U.S. has four separate sites, (in the States of Washington, Idaho, South Carolina and New York), that have been in the process of dealing with this 'nuclear weapons legacy' for the past several decades, and it is expected that the 'clean-up' will continue for at least another one or two decades (App E.3a, p 1 - 3). Ironically, the purpose of the West Valley Demonstration Project in New York State [Folga, 1996], was to study the management of H-L-W. It operated only from 1966 to 1972 and was closed permanently in 1976. In 1977 it became a legal requirement for the U.S. to consider all its 'spent fuel' as waste [Carter, 1997].

Nuclear Weapons Summary: As shown in Figure 4.4 (App E.3a), the anticipated eventual U.S. total H-L-W, from its 1943 until the 1980's 'weapons frenzy', and subsequent 'nuclear-weapons limitation Agreements with Russia' [Sittlow, 2020], is significant. Spent-fuel in immobilizing casks 34,800 t; H-L-W in 'Vitrified' or 'Calcined' form: 20,947 t. This 55,747 t is, naturally, awaiting some location for 'permanent disposal'. (App. E.5 for "Yucca Mountain Disposal Site" details). From 1945 to 2016 the World Total H-L-W from Nuclear Weapons Production: 315 Kt.   This 315 Kt does not take into account any material that may still be in 'Weapons Reactors' or have been diverted to 'civilian Reactors'; Therefore, it is likely that the 'true number' may actually be much larger.   However, this leads to about 315,000 tonnes that must be completely removed from our environment for many hundreds, thousands or even millions of years]!

4.2.4 Summary of High-Level waste: This chapter concludes with a summary of the disposal issue, by mass. By about 2018, a minimum of 315 Kt, ('Weapons') + 36.5 Kt, ('Vitrified') + 145 Kt, ('repackaged fuel') leads to a total of 496,000 tonnes of H-L-W requiring Permanent Disposal. This number completely ignores the 190,000 tonnes of 'spent fuel still in 'wet storage', as well as future tonnes estimated in the previous section.

Table 4.2, World Total of H-LNW

Type of Highly Radioactive Material	Storage Method	'World' Total Amount, (Kt)	Planned 'Future' for the Material
Weapons	Dry	315	Permanent Disposal
'Vitrified'*	Dry	36.5	Permanent Disposal
'Repackaged'	Dry	145	Permanent Disposal
'Spent-Fuel'	'Wet'	190	Unknown
Future 2019 - 30		32.8	Permanent Disposal
Future 2031 - 50		90	Permanent Disposal
Total		810,000 tonnes	

This leads to a total disposal need for H-L-W of close to 810,000 tonnes.

Figure 4.2,
Warhead
Inventory

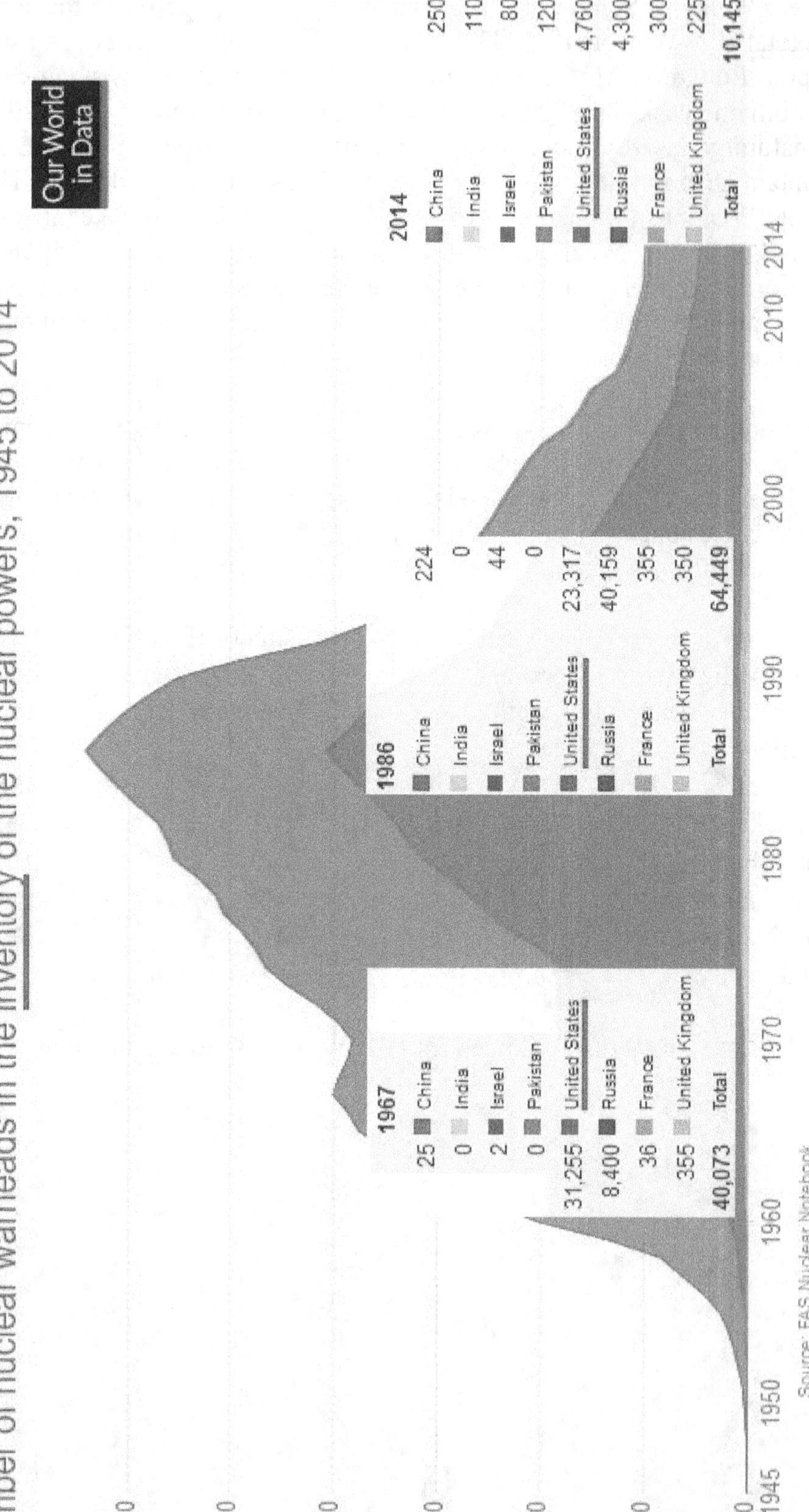

4.3 International Disposal Plans

Naturally, we want to dispose of the H-L-W in such a way that it no longer poses a threat to Earth and our way-of-life. Currently the consensus[] among 'nuclear Nations' is for "Deep Geological Repositories", DGR.   There are several interesting driving functions that are in play regarding this approach.

- If reprocessing is not chosen for the 'spent-fuel' from Reactors, then the time taken for the 'no-longer-wanted-fuel' to become 'only' as dangerous as natural uranium is approximately 500,000 years, (App E.2d).

- Therefore 'disposal' implies it must be extraordinarily unlikely for humans to encounter it accidentally, or purposefully, and that no known natural or Geologic processes will make such exposure possible!

- The U.S., to date, is the only Nation to have actually done a significant amount of construction on such a site, Yucca Mountain in Nevada, (App E.5). It was first proposed in 1979 and the first contract was signed in 1987 [sole repository]. Actual tunneling into the mountain started in 19942. However, the many delays since [fiasco, 2013], have made the future of Yucca Mountain as a DGR extremely uncertain at best.

- Finland[] is nearing completion of preparatory construction of its DGR [Finland DGR] and plans to apply for an operating license by 2021. It hopes to begin actual disposal operations in 2025. Finland's planning for all this started in 1983.

- From 2004 to 2014 Sweden constructed a research facility, 455m deep underneath the municipality of Oskarshamn. However, in 2011 Sweden selected Osthammar for the DGR site instead [towns compete]. Once again though, Sweden's DGR plan was stalled [DGR stalled 2018], in 2018 when a Swedish Court ruled that more information about the site's safety be ensured for the "thousands of years needed for the radiation levels to return to the original low levels of natural uranium "A processing plant to package the spent nuclear fuel and other H-L-W started construction in late 2019. An operating license permit will be applied for in 2021, depending on the Court decision cited above. From this it is estimated that actual 'burial' of the waste is unlikely to start before at least 2025.

- France, in 1999, selected Bure [French DGR], for its DGR site. Construction was expected to begin in 2022, with initial Waste disposal by perhaps 2030, but, opposition to the Project [anti-nukes 2018] appears to be intensifying.

- Russia, may also be getting a bit closer since it has selected a site for an underground Research Laboratory in the Krasnoyarsk Region, specifically the Nizhnekansky granitoid rock massif in Zheleznogorsk[] That may also become the site for a DGR if the initial results are satisfactory. Opening of the DGR is not planned to occur before 2030.

As far as could be determined (June 20, 2020), none of the other 46 Members [WNA members] listed have advanced even as 'far' as this with their plans for Deep Geological Repositories. If, as appears increasingly likely, DGRs will not become a widespread

---

2 [Yucca 1994]

reality in the very near future, then what options exist? There are no real alternatives to Deep Geological repositories on Earth.

4.4 Space Elevator Operations for High Level Radioactive Waste Disposal

The Transportation Story of the 21st Century will be the space elevator as it will revolutionize access to space. The website www.isec.org has a tremendous amount of information showing how and why the space elevator will become a significant contributor to the problem of nuclear waste disposal. The latest study report, done by ISEC, showed the throughput for the space elevator with a mature infrastructure would provide 170,000 metric tons of payload capability per year to GEO or the Apex Anchor. If the mature architecture was dedicated only to the nuclear waste disposal, it would take about three years to dispose of current and near term H-L-W in space and never threaten humans again.

Figure 4.3, Total Space Elevator Throughput [Swan, 2020]

As shown in other ISEC Studies, the Space Elevator should be able to operate as a routine, massive payload capable, safe, daily, and

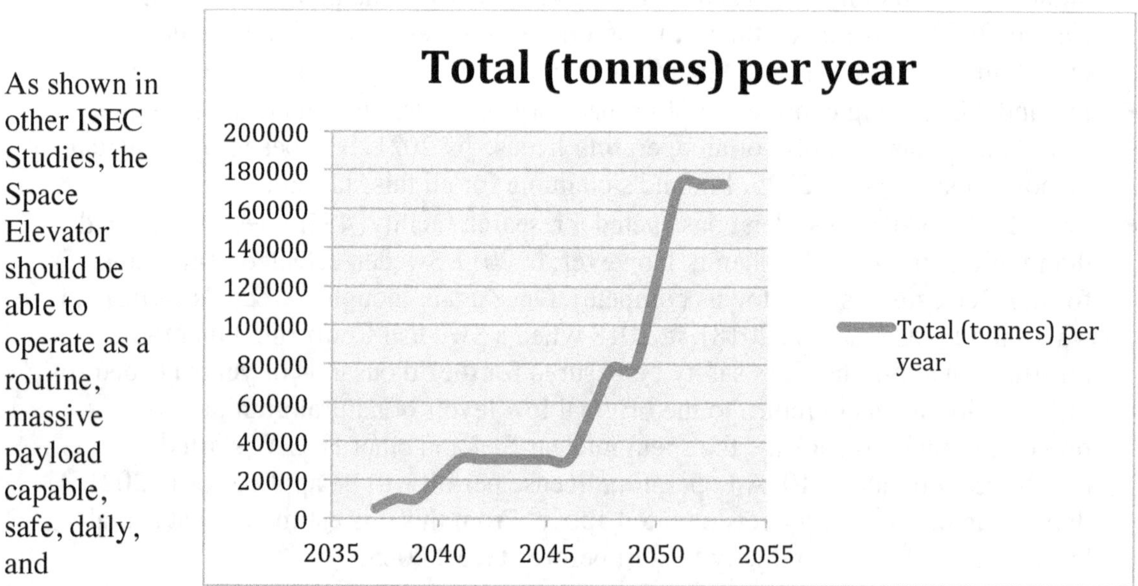

environmentally friendly approach to send the H-L-Waste beyond GEO. The throughput for this type of mission is shown in the following graph. It illustrates the growth once the program has been initiated.

The current design has the initial capability of 14 tonnes of payload per day per space elevator. This will grow to the carrying capability of 79 tonnes payload per space elevator as the system matures. In addition, there will be six space elevators distributed around the equator. When the high-level nuclear waste is packaged into canisters or containers of that mass or less, the concept is simple. Raise the radioactive waste to 100,000 km and release towards the Sun for permanent removal, never to see the Earth again. By starting it with a velocity of 7.76 km/sec, in the opposite direction of Earth's movement around the Sun, the package would fall into an elliptical orbit about the Sun within the Earth's orbital radius from the Sun and would have a period not tied to the Earth's 12 month cycle. With these two facts dominating the motion around the Sun, the disposed of high level nuclear waste would not come close to the Earth ever again - Conclusion: This would be a permanent solution for removal of H-L-W from our human habitat.

4.5 H-L-W Container Mass

To load the high-level waste onto the Tether Climbers, the size and mass must be compatible with the transportation system. The initial approach would be to use the vitrified/calcined waste that is currently in metal containers of roughly 2 cubic meters and mass of approximately 0.5 to 2.3 tonnes (filled and sealed).

There were 150,000 of these containers as of 2018. As this waste will be increasing over the development program of the Space Elevator, there will be additional containers waiting when operations are initiated. In addition, the dry spent fuel could be easily accommodated into canisters as well. There are many situations where the waste has been protected with a long-term storage approach using burial in concrete casks constructed from an inner steel cylinder containing the actual fuel-rod assemblies. There are between 5 and 10 thousand of the 5 to 10 tonne types of castes as well as 20,000 very large casks, approximately 100 tonnes each, that would have to be repackaged before being sent on the space elevators.

A good example of a problem is shown in the next image. These long-term storage casks are overwhelming the current approach for permanent disposal, but could easily be accommodated by space elevator tether climbers capable of 14 tonne payloads. These 3250 dry storage casks (US only), handling 31,000 tonnes of high level waste are about 9.5 tonnes each (fuel bundle only) and could easily be handled by the tether climbers.

Figure 4.4: Storage Casks

World Nuclear Association states: With repository development derailed, storage space at some operating nuclear reactors ran out, and at most of the 65 nuclear power plant sites (60 operating, in 2017) pool storage is being supplemented with dry cask storage. Of the total inventory of 78,600 tonnes of used fuel at 74 reactor sites in 35 states l, about 40% was in dry cask storage at the end of 2016. The total increases by 2000 to 2400 tonnes each year. In July 2015, 23,000 tonnes of used fuel was in dry storage at 67 sites, using 2159 dry storage casks (2463 casks at end of 2016). As of the end of 2019 the total inventory of used fuel was 84,000 tonnes. https://www.world-nuclear.org/information-library/country-profiles/countries-t-z/usa-nuclear-fuel-cycle.aspx.

Much of the 240 Kt would appear to require the 100t capability mentioned earlier, since the 'long-term storage methods often involve very large casks constructed from an inner steel cylinder containing the actual fuel-rod assemblies. This is surrounded by a thick, (~20 cm), layer of concrete that serves the dual-purpose of enormously reducing the amount of radiation that escapes, (App p 3) and also acts as the 'long-term' part, (~ 50 years or longer), of the barrier against corrosion from normal weathering. Since the climbers must already be radiation hardened, (to pass through the Van Allen Radiation -Belts41), the concrete can likely be removed before the H-L-W it begins its journey to Space and ultimate disposal! This would greatly reduce the total mass, perhaps to as little as 15 t, (9.5 t of SNF plus the thin steel 'liner').

4.6 Space Elevator Delivery to Safe Disposal Orbit:

When the Space Elevator is operational with a "train-like" schedule and routine departures from the Earth's surface climbing to space without rockets, safe removal of HLNW can be achieved. The approach would follow the following sequence:

- Starting Point: Earth's Orbit (149 million km circular)
- Destination Point: Disposal Orbit (much smaller orbit never approaching Earth) Approach (see next figure):
- Step One: Climb space elevator to 100,000 km altitude - gains energy and results in a 7.76 km/sec velocity at release.
- Step two: Release in negative direction (opposite the direction of Earth's rotation around the Sun) resulting in 21.94 km/sec velocity
- Step three: At perihelion a thrust is applied to reduce orbital energy (negative velocity direction to decrease the size of the resulting orbit)
- Step four: Refine orbital characteristics with remaining fuel, then prepare for long safe duration storage, such as venting fuel and reduce stored energy.

 These steps and procedures will ensure safe removal of high level nuclear waste to a disposal orbit never to reach the Earth again.

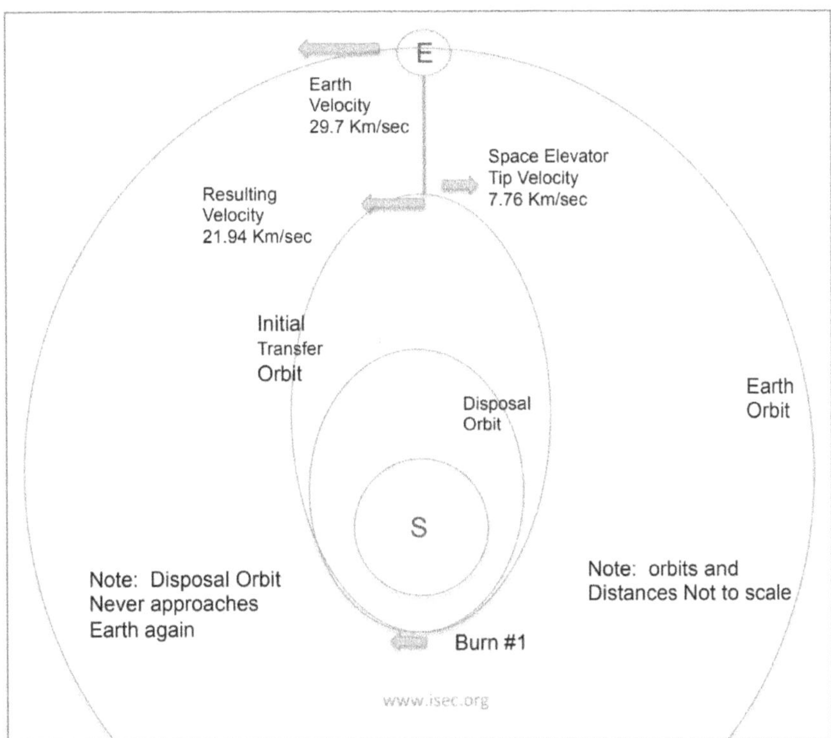

Figure 4.5, Sequence for Sun Earth Disposal Orbit

4.7 Summary and Conclusions:

The use of Nuclear Reactors to safely generate enormous amounts of inexpensive energy without adding to Greenhouse Gases is far too important to our Society to consider abandonment, at least until Space Based Solar Power constellation of satellites has matured. The hope of disposing of the hundreds of thousands of tonnes of highly radioactive waste by creating Deep Geological Repositories seems to be fading with each decade that passes, since not a single DGR is nearing 'operational status.' Finland, with its 'hoped-for-opening-by-2025,' is closest of all of the 51 Members of the World Nuclear Association3; In addition, the disposal of the radioactive waste from nuclear weapons development must be handled in a parallel manner. Permanent disposal is expected and can be achieved by putting it into a smaller solar orbit.

The only feasible alternative to DGRs is to toss the radioactive waste into SPACE so that it can be shown that the radioactive waste can no longer, even millennia in the future, pose any danger to humanity or our environment. Space Elevators, (once successfully constructed), will routinely carry massive amounts of cargo, including highly radioactive waste, to the Apex Anchors where it will impart high velocity towards an orbit around the Sun with no danger of "coming back to Earth."

Solving the high level nuclear waste problem would encourage nations to aggressively consider nuclear power for meeting their future electrical needs.

---

[3] [Fusion 2020]

Chapter 5 - Environmental Benefits of Space Elevator: Sun Earth L-1 Solar Shade

5.1 Introduction:

> Reducing the energy from the Sun that reaches the Earth's Atmosphere,
> could reduce the solar energy reaching the Earth by 1.8%. [Angel 2006]

With the tremendous concern for global warming, the idea of shading the Earth from the full brutal force of sunlight should be high up on the list of engineering solutions that could be undertaken. In the past, the idea of releasing 20 million tonnes of spacecraft towards the Sun lead to estimates of extremely high costs of launch for the program or seemingly impossible execution problems. With the motivation to significantly cool the Earth (approximately 1.8% cooling continually) and the refinement of the Space Elevator massive payload lift capability, the belief that Global Warming could be stopped must include this concept. It would be enabled by Space Elevators.

University of Arizona Professor Roger Angel wrote in his abstract to his remarkable paper on the topic of "cooling the Earth" with the initial words:
> "If it were to become apparent that dangerous changes in global climate were
> inevitable, despite greenhouse gas controls, active methods to cool the Earth on
> an emergency basis might be desirable. The concept considered here is to block
> 1.8% of the solar flux with a space sunshade orbited near the inner Lagrange
> point (L1), in-line between the Earth and sun."[4]

His concept, built upon a previous one from J. Early in 1989, proposed several ideas that could work within a 25-year period for "a few trillion dollars." When one takes into account the remarkable strengths of Space Elevators to place the needed spacecraft hardware in the proper location (Earth Sun L1), then the feasibility of the concept becomes real and could be accomplished in the near future. The following sections of this chapter describe the approach and the concept and then explains how the transportation issue becomes trivial - go to the Apex Anchor, assemble the large spacecraft and release in the negative velocity direction to slow down and "fall towards the Sun." (see later section on approach) Then on-board ion engines can direct the flight towards the Sun-Earth L-1 location. When the multiple spacecraft reach L-1, they release trillions of two-foot in diameter free flying small film solar satellites weighing one gram each.
During this description of the mission to cool the Earth Professor Angel defined his need at the L-1 location of 20 million tonnes of hardware to cool the Earth by 1.8 percent. Indeed, Space Elevators could enable the delivery of this amount of tiny spacecraft to the L-1 location ensuring a significant solar radiation management methodology. Once again, there is a customer "Demand Pull" that far exceeds the ability of rockets to supply and must wait for the operational Space Elevator.

---

[4] Angel, 2006b]

5.2 Greening Enhancement:

This chapter is structured to address the concerns of global warming and how a Space Elevator could enable an engineering solution. An Ad Astra Magazine article by Roger Angel and Pete Worden explained the huge concern that is reaching across the globe into all of our lives:

> "The Earth's surface temperature has risen by about 1 degree Fahrenheit in the past century, with accelerated warming during the past two decades. There is new and stronger evidence that most of the warming over the last 50 years is attributable to human activities. Increasing concentrations of greenhouse gases are likely to accelerate the rate of climate change. Scientists expect that the average global surface temperature could rise 1 to 4.5°F (0.6 to 2.5°C) in the next 50 years, and 2.2 to 10°F (1.4 to 5.8°C) in the next century, with significant regional variation. Global warming will have generally negative impacts on human life and the biosphere, so, to varying degrees, industry, scientists and policymakers are making significant efforts to mitigate the problem."[5]

5.3 Professor Angel's Approach:

To ensure accuracy on the approach for the development and deployment of the huge project, the next paragraphs are shared from "Space sunshade might be feasible in global warming emergency." [Angel 2006]

> "The possibility that global warming will trigger abrupt climate change is something people might not want to think about. But University of Arizona astronomer Roger Angel thinks about it. Angel, a University of Arizona Regents' Professor and one of the world's foremost minds in modern optics, directs the Steward Observatory Mirror Laboratory and the Center for Astronomical Adaptive Optics. He has won top honors for his many extraordinary conceptual ideas that have become practical engineering solutions for astronomy. Angel presented the idea at the National Academy of Sciences in April and won a NASA Institute for Advanced Concepts grant for further research in July. His collaborators on the grant are David Miller of the Massachusetts Institute of Technology, Nick Woolf of UA's Steward Observatory, and NASA Ames Research Center Director S. Pete Worden.

Angel is now publishing a first detailed, scholarly paper, "Feasibility of cooling the Earth with a cloud of small spacecraft near L1," in the Proceedings of the National Academy of Sciences. The plan would be to launch a constellation of trillions of small free-flying spacecraft a million miles above Earth into an orbit aligned with the sun, called the L-1 orbit. The spacecraft would form a long, cylindrical cloud with a diameter about half that of Earth, and about 10 times longer. About 10 percent of the sunlight passing through the 60,000-mile length of the cloud, pointing lengthwise between the Earth and the sun, would be diverted away from our planet. The effect would be to uniformly reduce sunlight by

---

[5] [Angel, 2006a]

about 2 percent over the entire planet, enough to balance the heating of a doubling of atmospheric carbon dioxide in Earth's atmosphere.

"Researchers have proposed various alternatives for cooling the planet, including aerosol scatterers in the Earth's atmosphere. The idea for a space shade at L1 to direct sunlight from Earth was first proposed by James Early of the Lawrence Livermore National Laboratory in 1989. "The earlier ideas were for bigger, heavier structures that would have needed manufacture and launch from the moon, which is pretty futuristic," Angel said. "I wanted to make the sunshade from small 'flyers,' small, light and extremely thin spacecraft that could be completely assembled and launched from Earth, in stacks of a million at a time. When they reached L1, they would be dealt off the stack into a cloud. There's nothing to assemble in space."

The lightweight flyers designed by Angel would be made of a transparent material pierced with small holes. Each would be two feet in diameter, 1/5000 of an inch thick and weigh about a gram, the same as a large butterfly. It would use "MEMS" technology mirrors as tiny sails that tilt to hold the flyer's position in the orbiting constellation. The flyer's transparency and steering mechanism prevent it from being blown away by radiation pressure. Radiation pressure is the pressure from the Sun's light itself. The total mass of all the flyers making up the space sunshade structure would be 20 million tons. At $10,000 a pound, conventional chemical rocket launch is prohibitively expensive. Angel proposes using a cheaper way developed by Sandia National Laboratories for electromagnetic space launches, which could bring cost down to as little as $20 a pound. Once propelled beyond Earth's atmosphere and gravity with electromagnetic launchers, the flyer stacks would be steered to L-1 orbit by solar-powered ion propulsion, a new method proven in space by the European Space Agency's SMART-1 Moon orbiter and NASA's Deep Space 1 probe.

"The concept builds on existing technologies," Angel said. "It seems feasible that it could be developed and deployed in about 25 years at a cost of a few trillion dollars. With care, the solar shade should last about 50 years. So the average cost is about $100 billion a year, or about two-tenths of one percent of the global domestic product." He added, "The sunshade is no substitute for developing renewable energy, the only permanent solution. A similar massive level of technological innovation and financial investment could ensure that. "But if the planet gets into an abrupt climate crisis that can only be fixed by cooling, it would be good to be ready with some shading solutions that have been worked out."[6]

Image from Angel's paper

---

[6] [UOA, 2006]

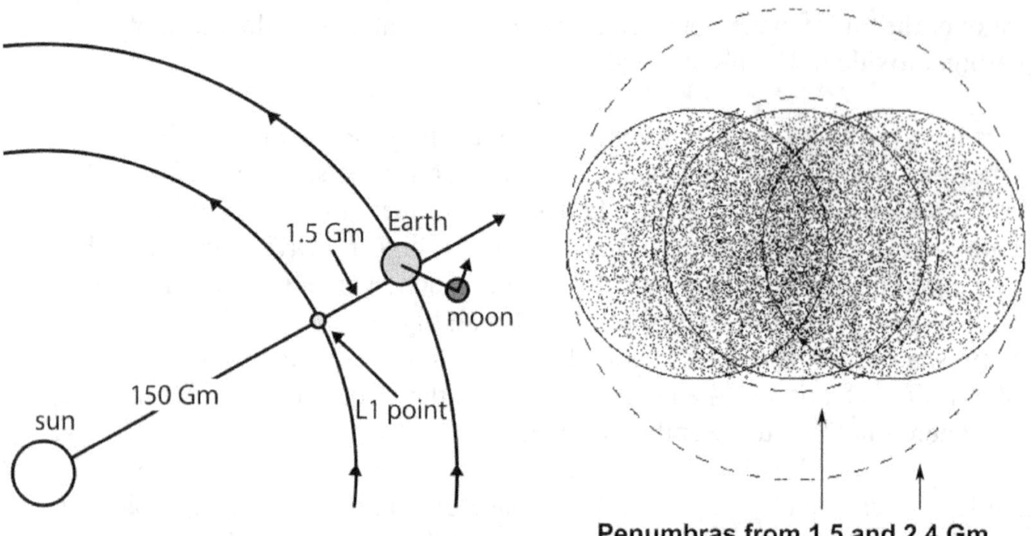

Fig. 5.1 Shadowing geometry. (Left) Schematic. The L1 point and the common Earth–Moon barycenter remain in-line as they both orbit the sun with a 1-year period (not to scale). (Right) Time-averaged view from Earth. The Earth wobbles with a 1-month period relative to the penumbral shadows cast from a sunshade at 1.5 and 2.4 Gm (dashed circles)[7].

5.4 Space Elevator Orbital Insertion Approach:

The strengths of Space Elevators change the approach for "how to get to space." The ability to move massive amounts of cargo to GEO and beyond opens up concepts that have been "shelved" in the past. This idea of cooling Earth by providing a shade at the Sun-Earth L-1 location is remarkable in both timing and achievability. The needs to address global warming are increasing from national to international levels of concern and potential actions. The key is that Space Elevators could implement this geo-engineering of the Earth's climate if the direction is given to accomplish this mission. There is nothing in Professor Angel's concept that could not be started now. This would enable trillions of two-foot diameter flyers to be available when the Space Elevator reaches its full operating capacity leading to 170,000 tonnes being placed at L-1. This approach could "enable this environmentally significant mission." The following steps would place the trillions of flyers in the L-1 location.

---

[7] [Angel, 2006b]

Sequence for New Location: Sun Earth L-1:

- Starting Point, Earth's Orbit (149 million km approx. circular)
- Destination Point, L-1 Orbit (1.5 million Km from Earth - orbital period matches Earth's)

Approach:

- Step One: climb space elevator tether to 100,000 km altitude
- Step two: Release in negative direction resulting in 21.94 km/sec velocity
- Step three: at perihelion thrust to reduce orbital energy
- Step four: at aphelion thrust to circularize orbit
- Step five: refine orbital characteristics and maintain orbit
- Step six: release the flyers

Figure 5.2, Orbital Approach for L-1 Solar Shield

5.5 Mass Movement:

As discussed in previous chapters, the only reasonable approach for moving massive amounts of tonnage to fulfill critically important missions in the near future is to avoid the rocket equation. Dr. Angel has stated he needs 20 million tonnes to be delivered to the Sun-Earth L-1 orbital position. If we calculated all the maneuvers to arrive at that rotating location, the

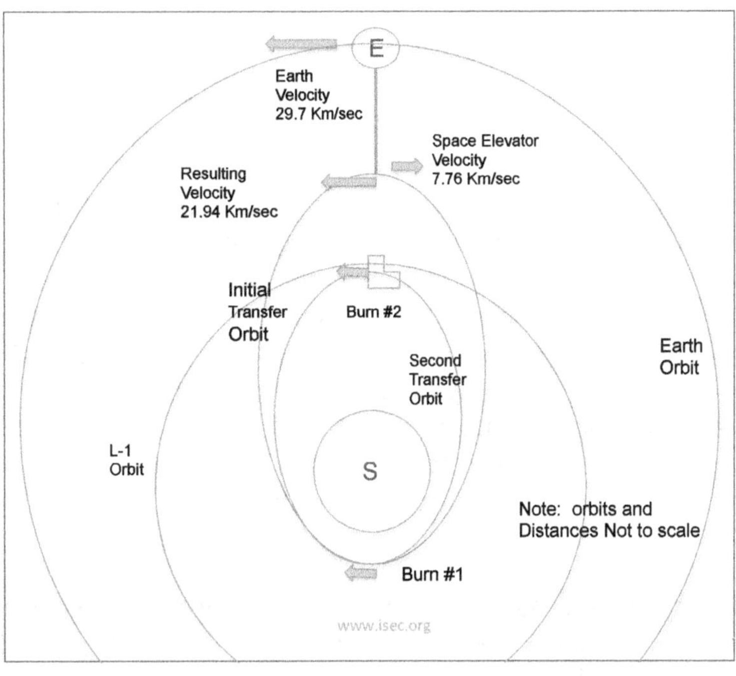

number of launches would be very large. This result is derived from the undesirable fact that the pad mass would have to be over 100 times the requested mass delivered to L-1. (percentage must be below 1%) As such, the need to deliver space systems to the location by other than rockets becomes paramount. The answer results in the need to launch 2,000,000,000 tonnes towards that location to have 20,000,000 tonnes arrive. This draconian fact is a "killer" for the program go ahead. But if you could use space elevators as a permanent train track to the destination, that would change the thinking drastically.

Reference Mission Sun Shade - 20,000,000 tonnes	Saturn V Rocket	Galactic Harbour Initial Operational Capability (2039)	Galactic Harbour Full Operational Capability (2045)	Galactic Harbour Robust Operational Capability (2052)
Throw Mass to L-1	45 tonnes	14 x 6 = 84 tonnes per day	79 x 6 = 474 tonnes per day	79 x 60 = 4740 tonnes per day
Launches Required	444,444	238,100	42,194	4,219
With Daily launches - How many years	1,218	652	115	11.5

Table 5.1 Galactic Harbour Fulfillment of L-1 Sun Shade Missions
Note, Saturn V escape mass - 45,000 kg - pad mass - 3,038,500 kg = 1.4%

The beautiful thing about a permanent infrastructure is that you could improve it and still have the same environmental effects from the operation with a tremendous increase in capability. The last column above is one that takes the current vision of the Space Elevator and increases it by ten. Most engineering solutions can be improved by ten by either building a factor of ten more sets of train tracks or by improving the actual operational through-put. I am sure no one living in 1830 would not have believed the amount of train traffic, or speeds seen in 2020, much less that crazy idea of flying in an airplane across the Atlantic. The last column above shows how the Sun Shade idea could be implemented with a robust statement of need (Demand Pull) and then an equally robust development of Space Elevators in both schedule and capability.

5.6 Conclusions:

Humanity needs to stop or turn back global warming. One concept that has much promise is providing a "sun shade" at the Sun-Earth L-1 orbital spot. This would block 1.8% of the sun's energy and help stop global warming. The critical element in this proposal has always been the expense for launch and the rocket equation mass delivery percentages. Once again, the old solution to this problem was to go to the Moon and develop a mining and manufacturing operation to make the trillions of flyers and then send them to the L1 spot, instead of manufacturing and delivering them directly from the Earth. Now that Space Elevators look real and should be operational by the second half of the 2030's decade, the potential solution to turn global warming around could come out of this "on-hold" idea. By ramping up Space Elevators with more tethers and bigger tether climbers, the movement of tonnage can go up linearly - exceeding 500,000 tonnes per year, enabling the sun shade idea to initiate operations within the early 40's decade. The realization is that: Space Elevators can Enable a Sun Shield for Earth at L-1.

Chapter 6 - Galactic Harbour Environmental Impacts

6.1 Introduction

The Space Elevator is a remarkable transportation infrastructure leveraging the rotation of the Earth to raise payloads from the Earth's surface into our solar system and beyond. It would indeed be part of a global transportation infrastructure. In a mature environment where Space Elevators are thriving in business and commerce, there would be several (probably up to six or more) spread around the equator, each with a capability of lifting greater than 14 metric tons of payload per day, Earth friendly and inexpensively. [Swan 2019a] From the beginning of this study, the authors have assumed and claimed that the Space Elevator would be environmentally friendly. We also felt that this assumption needed to be considered. and lead to the following conclusions

> Potential Beneficial Impact of Space Elevator: Daily operations, at zero (or negative) carbon footprint, reduces the environmental impact of the expected massive movement to space.

> Potential Beneficial Impact of Space Elevator: Reducing the number of rocket launches and replacing them with environmentally friendly climber lift-offs (such as to support humanity's movement off planet) will decrease pollution significantly.

This chapter explores the environmental impact of manufacturing, building and operating a Galactic Harbour Permanent Transportation System utilizing the Space Elevator construct. However, it must also be shown that the Space Elevator operations is virtually zero impact on the environment. This results from the fact that there are no effluents from the power usage nor impacts from the source of power (using solar cells to motorized wheels for driving a climber upward on a tether track).

6.2 System Overviews with Potential Environment Impacts

The following sections will have parallel items under each major segment of the Galactic Harbour. This will include a first quick description and then discussion on environmental impact of development and construction for each of the segments. The next question would be, what effect the operations and maintenance of the Galactic Harbour architecture would have on the Earth's environment. Taking each major segment separately, this chapter looks at the Earth Port, Tether Climber, Tether, GEO Node, and the Apex Anchor separately as to their possible impact on the Earth's environment.

6.2.1 Galactic Harbour: The Galactic Harbour will be the area encompassing the Earth Port of the Space Elevator [covering the ocean where incoming and outgoing ships/helicopters and airplanes operate] and would stretch, in a cylindrical shape upwards, to include tethers and other aspects outwards towards its Apex Anchors. The concept is that payloads come into the Galactic Harbour and are then processed before being released at some pier. The GEO Node is a good example of where a communications payload would be prepared for release, powered up, checked out, and then released towards its assigned slot at GEO. The intra-transportation system is very similar to a train operation, utilizing movement on rails from one station (Port or Pier) to another Customer product/payloads [satellites, resources, etc.] will enter the Galactic Harbour around the Earth Port and exit someplace up the tether [to LEO, to GEO region, to Mars, Moon, asteroids, intergalactic, towards the sun, dependent upon where it is released]. The current vision of a Galactic Harbour is shown in Figure 6.1. As a result, this Galactic Harbour will include structures far above the Earth

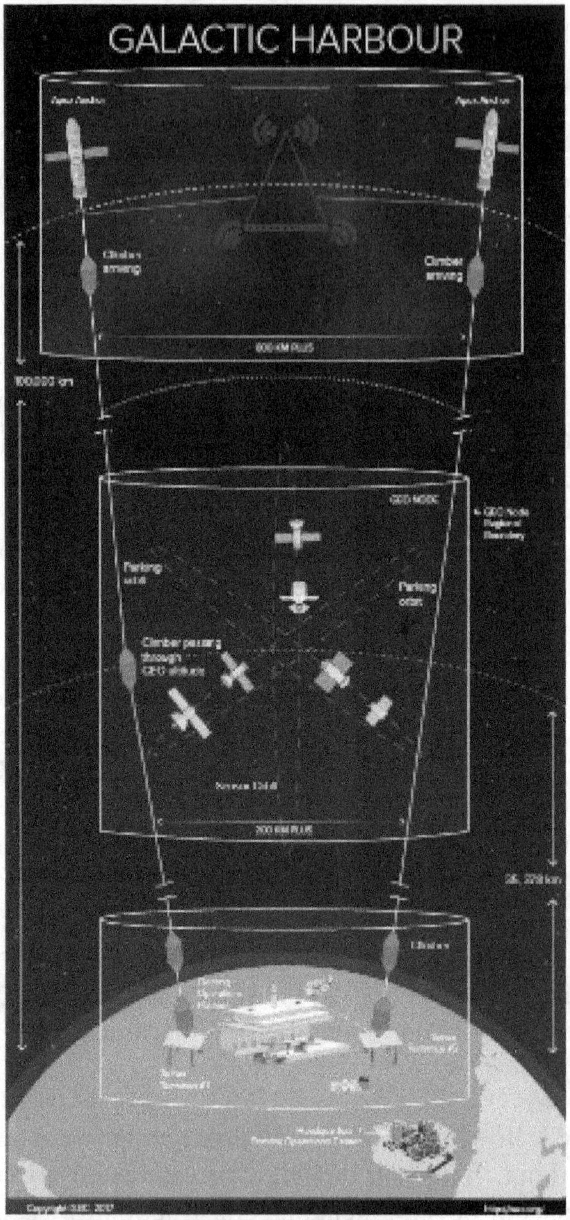

Figure 6.1: Overview of the Galactic Harbour Architecture [Swan, 2019a]

6.2.2   Tether: The length of the tether from Earth's surface is approximately 100,000 km, which would enable trips to Mars in as little time as 61 days. The current concept is a one meter wide, millimeters thick, tether with hundreds of individual layers of 100,000 km long single crystal graphene.

Tether Environment Impact: The tether would have the least impact on the environment within the Galactic Harbour Permanent Transportation System. At low altitudes, "bird strikes" in atmosphere would have to be considered,[21] as well as space debris at altitudes from the atmosphere up to the Apex Anchor. However, the tether is just a long band" so its most significant impact is if it "broke" and upwards of 2000 kms of tether would fall to the Earth. (very small probability - see two ISEC study reports at www.isec.org) The current thinking is to observe the fall of the tether as it "flutters" down into the atmosphere and lands on the surface of the ocean to the east of the Earth port where it would be reeled in as it fell. The environmental impact would be minimum as the tether is basically a carbon strip.

The manufacture of the tether (using graphene) could pose an environmental impact at the manufacturing site, but there are enough EPA regulations to minimize its effect on the local environment. However, the overall impact would be mitigated because the use of graphene could be considered sequestration of carbon and the organization would gain recognition for storing carbon after taking it out of the atmosphere.

[21]Notices would be provided to re-route airline traffic around the Earth port to minimize impacts to the Tether

6.2.3   Tether Climber: The word climber is used as the operative noun to denote the space system ascending or descending on the space elevator tether by its own means. Operational climbers are defined as the commercial version of a spacecraft taking customer payloads to altitudes such as LEO, GEO and Solar System trajectories. It will also return objects to disposal orbits or to the Earth's surface. The ascent power to climb would be electrical from the sun while the descent from GEO and ascent past GEO requires braking as centrifugal forces dominate. The variety of operational climbers will surprise even early believers in Space Elevators. There will be tether weavers, repairers, safety inspectors along with logistical trams, commercial climbers, human rated climbers, science climbers, hotels, and launch ports. An open standard will facilitate all manner of climbers to work on Space Elevators. The analogy would be the railroad's width of its rails. Anyone can put a train car on the rails if they adopt the standards. A similar approach must be used to ensure compatibility between tethers and climbers [Swan 2017b].

Tether Climber Environment Impact: The tether climber can be thought of as a self-contained vessel that would contain the payload going up to and returning from the Apex Anchor. Since it does pass through the atmosphere and ozone layer, the tether climber by itself would affect the environment in an insignificant manner. The design requirement for the climbers will require no environmental impacts from its construction or operations. It is the payload on the tether climber that would pose a potential risk to the environment and only if there is a mishap. Payloads such as fuel or nuclear waste would have to be in safe appropriate transportation canisters.

A modest significant environmental effect would be if the tether climber was dislodged from the tether and fell back to Earth. The tether climber would fall to the east of the Earth Port where there is over 2,000 km until reaching inhabited land. This would be no different than when issues occur on large oil platforms. The environmental impact would be any damage to the Earth Port and recovering the tether climber from the ocean. When something is dislodged at a higher altitude, the risk is still low as the resulting trajectory would be along the equator and would be designed to "burn up" upon entry to the atmosphere at great speeds - thus, very little impact on environment.

A matter to consider is where the tether climber would be manufactured and its impact on the environment during construction. Things like the ground system, ground water, local vegetation and effects on animal life would have to be considered around the manufacturing plant as in any environmental impact study.

6.2.4   Apex Anchor: A complex of activity is located at the end of the Space Elevator providing counterweight stability for the space elevator as a large end mass. Attached at the end of the tether will be a complex of Apex Anchor elements such as; reel-in/reel-out capability, thrusters to maintain stability, command and control elements, etc. [Fitzgerald 2017] The Apex Anchor mission is multi-dimensional; but its principal function is to provide stability for the Space Elevator as a large end mass. This will ensure a firm tether for the climber and provide a constant outward force. In addition, the Apex Anchor will have the mission of reeling the tether in and out as required for various tasks such as debris avoidance, damping tether end vibrations, and reacting to emergencies [Fitzgerald 2017]. One principal mission will be high speed releases of satellites into our solar system, such as to Mars in as few as 61 days [Swan 2020a].

Apex Anchor Environment Impact: The Apex Anchor is in space so its impact on the Earth's environment is minimal. Its greatest impact would be in the transportation of the materials from the Earth Port to assemble the Apex Anchor. This was discussed earlier as the materials would be the payload on the Tether Climber. Anything released at the Apex Anchor would be tossed beyond Mars and would never be dangerous again.

6.2.5   GEO Region: A complex of Space Elevator activities positioned in the Galactic Harbour's GEO region on the geosynchronous belt directly above the Earth Port. There will be several designated controlled volumes of space for various missions: one at each tether, one for a central main operating platform, one for each "parking lot of activity," and others. The GEO region is expected to become the centerpiece of a Space Port that provides "overhead" services such as repair/assembly of climbers, loading and off-loading supplies, servicing tugs and many other functions to a myriad of customers [Swan, 2017b] [Swan, 2017d].

GEO Region Environment Impact: The GEO region is in space so its impact on the Earth's environment is minimal. Its greatest impact would be in the transportation of the materials from the Earth Port to be assembled at the GEO region. This was discussed earlier as the materials would be the payload on the Tether Climber. In addition, the standard rules apply for space objects in orbits. They must be registered and maintained so as not to effect other objects. Indeed, the location inside the GEO segment would be important to independent spacecraft who run out of fuel, as they will be within the range of the GEO region's maintenance team and "tugboat." Indeed, the GEO region's operations would be "green" as it intends to clean up derelict satellites as part of its mission to increase safety in orbit.

6.2.6 Earth Port: The first Earth Port location is envisioned to be near the equator in the Pacific Ocean. Other locations on the equator are possible. The bottom line is that the Space Elevator will create a transportation infrastructure that will provide revolutionary access to space routinely, inexpensively, safely, daily and with large payloads. [Penny, 2015]

Figure 6.2: Space Elevator Earth Port [Image by lux Virtual
and Galactic Harbour Associates, Inc]

The Earth Port was formerly known as the Marine Node of a Space Elevator system before it became the focus of a remarkable transportation infrastructure. The Earth Port: [Swan 2017b]

- serves as a mechanical and dynamical termination of the Space Elevator tether, providing reel-in/reel-out capability and position management in order to deal with tension, wind and current
- serves as a port for receiving and sending ocean-going vessels (OGVs); provides landing pads for helicopters from the OGVs
- serves as a facility for attaching and detaching payloads to and from tether climbers and attaching and detaching climbers to and from the tether
- provides food and accommodation for crew members as well as power, desalinization and waste: essentially all things necessary for human habitation
- management, communications and other such support.

Earth Port Environment Impact: From an environmental point of view for the Galactic Harbour, the Earth Port poses a minimal but significant risk to the environment. The same regulations that govern offshore oil and gas platforms would be applicable to the Earth Port. Table 6-1 (at end of chapter) provides a summary of some examples of regulations pertaining to the protection of marine habitats and species in various regions around the world. The Earth Port would have some impacts regarding the surrounding area. These impacts would be similar to offshore oil and gas activities as shown in Table 6-2 (at end of chapter). Building the Earth Port would not pose any more of an impact to the environment as does the massive oil rigs located around the world. Transportation to and from the Earth Port by container ship or helicopter would also have a minimum impact on the environment. Currently, hundreds of container ships traverse the Earth's oceans with very minimal impact on the ocean's environment. Unlike rocket launches, the Earth Port will not affect the atmosphere or ozone layer. As discussed earlier, the Earth Port would have minimal impact on marine life around the platform. As a matter of fact, if the Earth Port followed the same procedures as cruise lines, the platform would provide food to the surrounding marine life (provided the discarded food is processed correctly).

When considering the Earth Port, located in the international waters in the Pacific (estimate 2,000 km west of Galapagos) the team must consider the Environmental Protection Agency's guidelines. The list of them are as follows:

According to the Environmental Protection Agency, the following twelve items must be addressed for something residing on the Ocean surface: [Cordes 2016]
>   1. Ecological Resources
>   2. Cultural/Native American Resources
>   3. Hazardous Materials/Waste Management
>   4. Land Use
>   5. Visual Resources
>   6. Noise
>   7. Geology and Soils
>   8. Natural Resources and Energy Supply
>   9. Traffic and Transportation
>   10. Water Resources (including Wetlands and Wild and Scenic Rivers)
>   11. Airspace
>   12. Environmental Justice

For each of these 12 items the environmental consequences considered for the site for a Space Elevator's Earth Port planned to be a floating platform in the ocean at a fixed location. There are several sites near the equator where these conditions prevail. This makes an Environmental impact study much easier than if the Earth Port was to be land-based. Of the 12 items required in an Environmental Impact study for the Space Elevator only a few require consideration:

>   3. Hazardous Materials/Waste Management - Hazardous
>   Materials/ Waste Management: materials such as nuclear waste containers would be handled only as to take them as delivered and placing them robotically on to climbers. Waste management would be handled as has been done on other ocean platforms for years.
>   8. Natural Resources and Energy Supply - No local natural resources would be involved; energy supply would be entirely local solar.
>   9. Traffic and Transportation- All traffic would be by ships and planes. Facilities would be established for container ships to off load and adequate space would be provided for material storage waiting for loading on climbers Any air pollution would not be any more then at a small sea port or airport.

10. Water Resources (including Wetlands and Wild and Scenic Rivers) - Fresh water would be provided by a desalination plant powered by solar power.

11. Airspace - Airspace would have to be carefully controlled to avoid the tethers and regulate landings and takeoffs, mainly by helicopters. Seaplane facilities may also be included. Refueling facilities for all craft both sea and air would be handled by standard procedures

As the Earth Port development activities would include these many considerations above, some recommendations, similar to placement of oil rigs in the open ocean, have surfaced already.

Recommendations: [Cordes 2016]
1. Establish robust baseline ecological survey data within planning area and in appropriate reference areas
2. Determine the locations, size and type of ecological and biological significant areas (EBSAs) through comprehensive surveys including visual imagery
3. Establish protected areas around significant areas of representative communities
4. Establish borders of protected areas to be set-back distances based on typical distances of impacts from installations:
   a. 200 m from seafloor infrastructure with no expected discharges
   b. 2 km from any discharge points and surface infrastructure
5. Consider activity and temporal management to restrict impacts
6. Implement a comprehensive and robust monitoring program that can reliably detect significant environmental changes in areas of exploration activity, areas inside the established MPAs, and reference sites outside of MPAs and activity zones.

As a result, the Earth Port consisting of one large platform or several connected smaller platforms would be a very "green" community. In total, the development of the Galactic Harbour would be very close to carbon neutral.

6.3 Green Operations of Galactic Harbours

Given the major sections of the Galactic Harbour permanent transportation system (Earth Port, Tether, Tether Climber, GEO Node and the Apex Anchor) are operationally environmentally friendly, the only section that would have a small impact on the Earth's environment would be the Earth Port, but very minimally. However, its impact would be no greater than the building and operating of large oil platforms currently being used around the Earth, without the major risk of major oil leakage from underwater drilling. [Penny 2015] The bottom line is that the Space Elevator will create a transportation infrastructure that will provide revolutionary access to space routinely, inexpensively, safely, daily, and with large payloads. It's design, construction and operations from the start will be environmentally friendly. [Penny 2015] In point of fact, the operations of a space elevator will be carbon negative. Several significant operational insights must be expanded upon to explain how significant the establishment of this permanent transportation infrastructure would be to the Earth's environment. Several of these concepts can be considered key elements establishing the reality that Space Elevators can make the Earth Greener:

1. Routine manufacture of graphene tethers will sequester carbon dioxide from the atmosphere. This would cross over from the original tethers to the future improvements of tethers and to the multiplication of tethers as other Galactic Harbours are created beyond the initial one.
2. Enabling Space Solar Power would by itself eliminate hundreds of coal power plants around the world. One questions could be: what would it take to completely replace fossil fuel burning plants that power the Earth today? See chapter3
3. Another massively positive environmental move would be to enable permanent disposal of high level nuclear waste towards the Sun. See chapter 4
4. Movement of Earth polluting industries into space locations while using materials from other bodies in space instead of from the Earth.
5. Interplanetary travel with minimal chemical-based propulsion, release at high velocity daily from the Apex Anchor opens up the solar system to scientists and explorers.
6. Assisting the removal of space junk will help limit additional debris as well as ensure safer travel in orbit.
7. Recycle, repair and refuel old satellites will require fewer rocket launches while increasing our presence in space
8. Practical logistical support for Mars or Moon colonies, without chemical rockets launching from Earth.
9. Placement of Planetary Defense sites at Apex Anchors could ensure Earth's protection without needs for rocket launches.

6.4 Conclusions:

In point of fact, the operations of Space Elevators and Galactic Harbours will be carbon negative. This chapter has shown:

> Potential Beneficial Impact of Space Elevator: Daily operations, at zero (or negative) carbon footprint, reduces the environmental impact of the expected massive movement to space.

> Potential Beneficial Impact of Space Elevator: Reducing the number of rocket launches with environmentally friendly climber lift-offs (such as to support humanity's movement off planet) will decrease pollution significantly.

Table 6-1: Some Regulations Pertaining to Protection of Marine Habitats and Species. [6]

Jurisdiction	What is protected	Implementation of Protection	Assessment and Monitoring
Australia	Sensitive features and values of the environment, particularly the presence of threatened species	Site-specific environmental plans developed by operators and vetted by commonwealth authority	Each activity requires an environmental plan approved by legislator; details not prescribed
Barbados	Some coral reefs and fisheries that fit conservation priorities	MPAs, small MPAs in place in coastal habitats	EIAs required, monitoring for emissions, discharges, biological indicators, 5-yr review cycle
Brazil	Cold-water corals	Designation as conservation unit	Monitoring of water, sediments, and biota required but methods not stipulated
Canada	Listed species, cold-water corals, unique/diverse/productive habitats	MPA designation, Areas of Interest, Sensitive Benthic Areas, Fishery closures, Marine Parks, Species-at-risk	Monitoring encouraged for exploration, mitigation plans and monitoring required for production.
Colombia	Coastal and marine areas that fit conservation objectives	National Natural Parks System, regional Districts of Integrated Management, Regional Natural Parks	EIA required, monitoring required, but methods not stipulated
Grenada	Coastal reefs, offshore fisheries, pollution of offshore areas prohibited	Benthic-Protection Areas (fisheries), MPAs (coastal habitats)	Required but not described
Israel	Unique habitats, high species richness, rare species, archeological sites	Proposal for establishment of MPA system, considering 600m dry-back distance	Strategic environmental survey required within 2 km, sediment sampling throughout, 8 video surveys within 500m
Jamaica	Coastal coral reefs, some offshore fisheries, discharge of "poisonous, noxious, or polluting matter" is prohibited	MPAs, Marine Parks, some in place in shallow waters	Baseline surveys completed, but not explicitly required
Malaysia	Fisheries and habitat quality, CITES listed species		EIA carried out by registered consultants, evaluation of impacts in accordance with international standards
Mozambique	No specific protections outlined. Rules for avoiding impacts and preventing deposition of toxic substances in the ocean		EIA is required

New Zealand	Sensitive environments and threatened species	MPA system in development, currently avoidance or mitigation	Baseline surveys for EIA only
Nigeria		No specific marine protections, but signatory on various international agreements	
Norway	Valuable and vulnerable areas, fisheries, sensitive species (e g, corals)	Currently defining a framework for oil and gas activities within Norwegian Climate and Pollution Agency	Baseline surveys required, monitoring required after production, monitoring includes fish condition, and benthic habitat condition assessments every 3-yrs
Portugal	Habitats and Species listed in EU Habitats Directive	System of MPAs. Existing leases in all mainland EEZ, no exploitation yet	During exploration phase, all measures should be taken to prevent pollution; EIA is only required for the exploitation phase
Tanzania		Legislation stipulating that "Environmental protections should follow best practices of industry"	
Trinidad and Tobago	Sensitive areas and sensitive species	MPA system being developed, one currently for shallow water reefs	Baseline surveys for EIA, monitoring endorsed but not required
UK	Habitats and Species listed in EU Birds and Habitats Directive, OPSPAR Convention, and other national conservation legislation	Network of MPAs with designation as Special Area of Conservation, Nature Conservation MPAs, and Marine Conservation Zones	Baseline surveys for EIA, monitoring endorsed but not required
US	High density biological communities	BOEM Notice to lessees, National Monuments, National Marine Sanctuaries	Mitigation areas determined from seismic anomalies. Visual surveys only required if near known high density communities. No monitoring required

MPA – Marine Protected Area, BOEM – Bureau of Ocean Energy Management, EEZ – Exclusive Economic Zone, EIA – Environmental Impact Assessment, NTL – No time Limit

Table 6-2: Concerns of Offshore Oil and Gas Platforms on the Environment. [6]
Types of impacts from offshore oil and gas activities

Concern	Nature	Extent	Environmental Issues
Drilling discharges (cuttings, drilling fluids, cement, chemicals)	Physical (excess sedimentation); chemical (toxic effects; enrichment effects)	100-500 m (solids) "Local"	Smothering; clogging of feeding and gas exchange structures; direct toxicity; altered electrochemical environment; changes in nutrient availability, decreased species abundance, altered community structure
Produced Water	Chemical (toxic effect)	1-2 km (produced water and dissolved components) "Widespread"	Direct toxicity; food-web contamination; potential food-chain; and trophic amplification
Anchors	Physical (direct damage; hard substratum)	"Local"	Direct physical impact at emplacement, potentially continuing impact through tidally induced motions; provision of hard substratum, for colonization by sessile epifauna and associates
Flow and control lines, umbilical's	Physical (direct damage; hard substratum)	"Local"	Direct physical impact at emplacement; increased sedimentation; provision of hard substratum for colonization by sessile epifauna and associates
Export pipelines	Physical (direct damage; hard substratum)	"Widespread"	Potentially extensive direct physical impact at emplacement; provision of hard substratum for colonization by sessile epifauna and associates
Risers	Physical (hard substratum in water column)	"Local"	Provision of hard substratum for colonization by sessile epifauna and associates
Anchors and pipelines	Direct physical disturbance	15m (direct impacts), 50-100m (indirect impacts)	Mortality and burial of benthic fauna; fragmentation of corals; increased sedimentation; pipelines can corrode; and increase toxicity
Surface structures and vessels	Restricted movement of vessels	Right-of-way for working vessels; 1-	Restricted industrial and scientific activity

		2 km for surface infrastructure	
Seabed infrastructure	Artificial habitat	Direct for sessile species, ~500m for pelagic species, potentially altering distribution over large areas	Altered distribution; may increase species connectivity (including invasive species)
Artificial light	Physical (energy, electromagnetic spectrum)	100s of m	Surface light attracts some mobile species and repels others; subsurface light impacts are largely unknown
Acoustic energy	Physical (energy, hydrostatic pressure)	200-300m (high intensity); up to 4000km (lower intensity); highly variable	Localized auditory damage (100ws of m), disruption of marine mammal behavior, and physiological stress; impacts to fish unknown; invertebrate larval impacts
Mass hydrocarbon release (atmosphere, sea-surface, water column seafloor)	Release of oil and gas; potential disposition of equipment; potential additional affects of mitigation efforts (dispersants, burning, etc.	Localized deposition of gear; pelagic and benthic impacts depend on size of event, ranging from 100s of m to 100s of km in diameter	Increased toxicity; altered benthic, pelagic, and infauna communities; mortality of corals
Mass dispersant release	Chemical (toxic effects), synergistic with oil exposure	Variable, depending on size of event	Increase toxicity; changes in microbial community; reduced settlement of larvae; mortality of corals

Chapter 7 - Beneficial Environmental Impacts from Dual Space Access Architecture

7.1 Introduction

Potential Beneficial Impact of Space Elevator: Reducing the number of launches (such as to support humanity's movement off planet) will decrease pollution significantly.

Potential Beneficial Impact of Space Elevator: Provides safe, reliable, routine, daily, environmentally friendly, and inexpensive transportation infrastructure to move massive tonnage to GEO and beyond, specifically the Moon and Mars.

When we look at the Moon and think of the Apollo missions, we forget how extremely difficult they were to accomplish, both in energy and design complexity. Tsiolkovsky's remarkable rocket equation consumes so much mass to achieve orbit that, historically, we have been restricted as to what we can deliver. Now that we have decided to go to the Moon and on to Mars in a combined international, government and commercial effort of great magnitude, we need to expand our vision of 'how to do it.' It would seem that the establishment of a more robust infrastructure with reusable rockets and permanent Space Elevators must be developed. In this chapter, we will discuss the strengths and weaknesses of the components of this combined architecture with the purpose of placing mission equipment and people where they need to go and when they need to be there. The multiple destinations, complexity of orbits, magnitude of each transition to orbit, and infrequent launches currently means that the difficulty of fulfilling the dreams of the many is a monumental "reach." Expanding space access architectures to include Space Elevators will enable a robust movement off-planet. To quantify the understanding of the magnitude of the effort to move off-planet, the chapter will focus upon a Dual Space Access Architecture with a robust SpaceX Starship in operation (as an example of the future rocket capabilities being planned by many companies and countries - Blue Origin, SpaceX, NASA, China, ESA, and India) along with a mature six Space Elevator Galactic Harbour complex.

During the discussions for this chapter, the authors considered the strengths of rocket launches along with their difficulties. We recognize there are three major strengths: 1) rockets are successful today and great strides are forecast for the future, 2) reaching any orbit can be achieved, and, 3) rapid movement through radiation belts for people enables flights to the Moon and Mars. In this chapter, we will also point out the strengths of a permanent infrastructure with daily, routine, environmentally friendly and inexpensive attributes. These Space Elevator strengths will be compared to the difficulties of executing a Space Elevator developmental program. It is revolutionary in that it changes the equation of delivery dynamics — rocket equation limiting vs. massive movement of logistical cargo — while it fulfills the customer's needs. It has so many promises, and is seen as an enabler for so many dreams, that it must be pursued, now. Space Elevators will not be ready for the initial human migration off-planet. However, once colonies are established on the Moon and Mars using rockets, Space Elevators will enable robust growth of the colonies by moving massive cargo, daily, inexpensively, environmentally friendly, and routinely.

7.2 Customer Demand Pull

As said: "Good ideas are good ideas - ONLY - if you have a passionate advocate for that idea." In two arenas there are vocal proponents of their visions and their needs for the future beyond Low Earth Orbit -- GEO for Space Solar Power and Mars (or L-5) for space colonies. Each of these have tremendous visions that say "let's go do this, as it is important and timely." Of course, that leads to customer demands to be fulfilled by a transportation infrastructure that will be in place to support their visions and designs. Who will supply the tonnage to far away places to accomplish these visions?

Space Solar Power (initially discussed in Chapter 3): In the space solar power arena - and there are many proponents around the world - most try to ensure that their research and experimentation are supported by Dr. John Mankins. He has taken the concepts of Dr. Peter Glasser and has developed them into realizable engineering concepts that can achieve a monumental task. Dr. Mankins has stated that he believes a robust space solar power architecture would "stop global warming, and even reverse it."[8] His goal is to satisfy 12% of global electrical power demand by 2060 with energy from Geosynchronous orbit. His objective is to eliminate 100's of fossil fuel burning plants. He needs five million tonnes[9] of operating satellites (acres in size) delivered to 36,000 km altitude. This is a remarkable mass when you understand the rocket equation and realize that to accomplish this goal, requires 250 million tonnes on the launch pad at liftoff (2% to GEO). Mr. Musk's estimate of 21 tonnes to GEO (for his Starship booster without refueling) would require over 238,095 launches - even at three per day, that is 217 years.

This demonstrates demand pull for the Space Elevator. To reach Dr. Mankins' goal, Space Elevators should be a part of his implementation approach for his project. When Space Elevators are a mature infrastructure (six Space Elevators around the equator - three Galactic Harbours with a capacity of 79 tonnes per day per elevator - say 2045?) they will provide 173,010 tonnes to GEO per year. This leads to 29 years during Dr. Mankins' development time.

Colonies Off-Planet (Mars and L-5): There are two concepts that have been greatly discussed: Elon Musk's Mars Colony and Dr. Gerard O'Neill's L-5 rotating cylinder. Here are a few descriptive sentences and a recognition of what they need for their massive movement of cargo.

Mr. Musk's Mars settlement has been discussed extensively ever since he proposed a rocket design that could provide massive movement towards Mars. His design of the Starship

---

[8] [Mankins, 2020]

[9] [Mankins, 2019]

vehicle shows great hopes for success. In addition to suggesting that he will have a colony with upwards of a million settlers in place within his lifetime, he has stated that he requires 1,000,000 tonnes of cargo to support them on Mars. This customer need is straightforward. His images are well laid out and his understanding of the problem is excellent. As such, his demand pull for a transportation infrastructure is 1 million tonnes delivered during the decades of 2020 through 2060.

An older dream of the National Space Society (carried over from the original L-5 Society) is to create a working and living space at the stable location trailing the Moon at an equivalent radius from the Earth. This Earth-Moon Lagrangian Point (L-5) orbital spot will provide a very suitable location for placing a million people inside a rotating cylinder. The dreams, visions, images and engineering designs have been around since the mid 1970s. However, the ability to deliver 11 million tons of supplies, infrastructure, power, water, oxygen and fuel was just non-existent. As such, the dream has been "out there" but not fulfilled. Once again, an unsatisfied customer demand.

Table 7.1, Delivery Statistics for Customer Demands

Vision	Space Solar Power at GEO	Mars Colony (SpaceX)	L-5 Colony
Customer Demands Tonnes to be delivered	5,000,000	1,000,000	11,000,000
Years to deliver by mature Space Elevators at 170,000 tonnes / year	30	6	65
# Launches - Space Launch System (to TLI 45 tonnes)	111,111	22,222	244,444
# Launches - Starships (to TLI 21 tonnes)	238,095	47,619	523,809
# Launches - New Glenn (to TLI 20 tonnes)	250,000	50,000	550,000

(note: For Starship to deliver 100 tonnes to GEO, or Moon/Mars, there must be multiple support launches - refueling/supply/crew)

To put this set of numbers in perspective, the following chart was used in the International Academy Astronautics report in 2014 [Swan 2014] to project needs of the customers.

Table 7.2, Customer Demands (tonnes per year) [Swan 2014]

Demand in Metric Tons	2031	2035	2040	2045
Space Solar Power	40,000	70,000	100,000	130,000
Nuclear Materials Disposal	12,000	18,000	24,000	30,000
Asteroid Mining	1,000	2,000	3,000	5,000
Interplanetary Flights	100	200	300	350
Innovative Missions to GEO	347	365	389	400
Colonization of Solar System	50	200	1,000	5,000
Marketing & Advertising	15	30	50	100
Sun Shades at L-1	5,000	10,000	5,000	3,000
Current GEO satellites + LEOs	347	365	389	400
Total Metric Tons per Year	58,859	101,160	134,128	174,250

Summary of Demand Pull: This is when a future customer needs something and asks the developer to supply it. A customer needs a capability by a certain time and will help to mature a technology and stimulate the start of a supporting element of its program with the offer to be an "anchor tenant." ISEC needs to recognize these critical customers' demands and then gain support from them to move forward. The report shows two demand pull opportunities: Dr. Mankins needs five million tonnes to Geosynchronous. In addition, these two settlements just discussed have been dreamed of and are now closer to fulfillment for two reasons. The Human species has decided to move off planet making it a dream of many instead of the few; and, a Space Elevator is very close to being started - enabling the construction, supplying and staffing of these colonies.   At 170,000 tonnes per year delivery to GEO and beyond, these customers can be supplied. If one was to look at the huge needs from these two programs, as well as the programs in the table above, it would seem they should now be demanding a Space Elevator capability as soon as possible.

7.3 Dual Space Access Architecture

The essence of the Dual Space Access Architecture is that the two methods of achieving spaceflight are complementary and compatible rather than competitive. Each has its own strengths and weaknesses. Future rockets are being designed now to deliver payloads to the Moon in the near term (2023 +). Next comes their growth to massive launches, in both number and payloads, occurring in the second half of this decade with the 30's having mature rocket architectures. Development on the Moon, and initial colony on Mars, will be well along by the early 30's. By the end of that decade, Space Elevator infrastructures will be incrementally built with more and more capability leading to many complementary missions. Our vision towards the turn of that decade, starting in 2040, is that there will be six Space Elevators located around the equator helping with the delivery of massive amounts of payloads to GEO, the Moon, and Mars. Our concept will have daily departures to each, rapid travel to each (as fast as 61days to Mars), and with massive amounts of payload to support people at multiple destinations. A good way to express the team arrangement could be: Rockets to Open up the Moon and Mars with Space Elevators to supply, grow and insure the sustainability of the colonies.

The comfort level is high and the ability to repeat is well known with rockets. As one who lived through the marvelous development of human spaceflight, this capability to move humans with rockets is an exciting strength. Trips to the Moon highlighted the commitment of moving off planet with remarkable flights pushing the envelope of risk and achievement. Once a Space Elevator has been installed and upgraded to its initial capability, it will be there for a very long time, similar to roads, bridges or train tracks. Rising from the surface of the ocean to the Apex Anchor (100,000 km as a starting concept) is accomplished with external power - such as solar energy. The strengths of this Dual Space Access Architecture will enable human migration off-planet robustly and safely. Space delivery can become as routine as UPS, Fed-Ex, Amazon, and DHL are today. One significant conclusion is that using the strengths of both parts of this Architecture enables so much more than the individual parts. Customers must also take into account the short-falls of each major segment of the Architecture.

7.4 Avoiding the Rocket Equation:

As of 2020, rocket systems are working towards many improvements that will have significant impacts. The reusability of major segments will lower the cost while making the rockets massive will enable bigger payloads. However, the major flaw of the rocket approach is the consumption of its initial mass at the pad to increase the velocity required for flight. This consumption of pad mass is a huge portion of the total vehicle weight and decreases the payload capability for each launch. Essentially, to reach LEO, the rocket equation consumes 96% of pad mass (fuel burned and structure used). The remaining 4% is the payload mission equipment, with everything else released earlier for reuse, left in lower orbits as debris, or burned up to gain velocity. The reality is that 17,000 miles an hour to stay in LEO is demanding. Then, to gain velocity to go to the Moon, GEO, or Mars, the rocket equation demands consumption of more fuel, structure, electronics and equipment along the way. The final velocity is hard to reach and those other parts of the rocket do not contribute to the next stage of the mission; and so, they are "thrown-away" while consuming fuel. Only two percent is sent towards high orbits enabling their escape from low Earth orbit. I like to point out that the real catastrophic number illustrating this point was that the Apollo equipment that landed on the Moon (with Astronauts) represented less than half a percent of the mass on the launch pad at Cape Canaveral (only 0.17% returned to Earth). Consuming fuel, structure and equipment to gain velocity is a brutal approach - of course, it is the only approach today; but, it is still brutal. It should also be pointed out that there are no "cost" or "reusability" factors in the rocket equation. You can do it more efficiently, but you cannot beat the 123 year old equation. (note: Estimates with more reusability in rockets lowers the delivery percentages)

7.5 An Example of Dual Space Access Architecture - to Mars

This study will lay out the current SpaceX approach with Starships and compare that to the movement of cargo to Mars by a robust Galactic Harbour. Price per tonne is variable and cannot be estimated at this point, but the movement of massive cargos required can be compared. The timelines are essentially:
- Late 2020's to Early 2030's: Starships to Mars with people
- Late 2030's to Early 2040's: Mature Galactic Harbours

Vision of the Future: On to Mars. If the Mars access strategy has two components, you end up with a much stronger position having both advanced rocket architectures and permanent Space Elevator infrastructures supporting movement off-planet. Assessing the previous discussions showing strengths and weaknesses, the logical conclusion is that there should be a concerted effort to ensure development towards a Dual Space Access Architecture. Some basic realizations are that:

>   (1) rockets should be emphasized for people movement,
>   (2) rockets have tremendous strengths for LEO/MEO destinations,
>   (3) Space Elevators should be leveraged for GEO and beyond, and
>   (4) Rockets should be leveraged to open up colonies on the Moon and Mars
>   (5) Space Elevators should be leveraged to deliver cargo, equipment, and supplies for Lunar and interplanetary missions.
>   (6) Space Elevators should be leveraged to build up the colonies.

These concepts lead to the realization that developmental planning must be initiated in the very near future for both advanced rockets and Space Elevators. The strategy is to use Rockets to Open up the Moon and Mars while leveraging Space Elevators to supply and grow the colonies. With this as a baseline for the development of the Mars Colony, the next few sections will show an analysis comparing the mass delivered and time to travel to/from Mars using the "Starship" or the "Galactic Harbour Permanent Transportation Infrastructure."

Table 7.3 Percentage to Destination

Launch Vehicle	Mass on Pad (kg)	Mass Delivery	%
Apollo Saturn V	3,233,256	Lunar lander = 15,103	0.5
		ocean landing = 5,557	0.17
Atlas V	590,000	to GEO = 8,700	1.5
Falcon Heavy	1,420,788	to GEO = 26,700	1.9
Starship	4,000,000	to GEO = 21000	0.5
New Glenn	1,323,529	to GEO = 13,000	1

7.5.1   SpaceX's Starship Approach to Open up a Mars Colony[10]:
Mr. Musk's desires are to have his colony of over 1,000,000 people on Mars during his lifetime (so let's give it 40 more years from 2020). In addition, he has stated that he needs

---

[10] This explanation is Dr. Swan's approach to understand SpaceX's approach: It is probably "off" target in a few points, but it is used as an example for understanding the Dual Space Access Concept.

one million tonnes delivered to Mars to support his developing colony. One of the first planning thoughts must be that the window to launch towards Mars from Earth, via a Starship, is only an 8-week window every 26-months. Once he has established the process for colonization and initiated the Colony with explorers and early settlers, SpaceX envisions loading the 1,000 Starships during the periodic 26-month window. Then all 1,000 (carrying approximately 100,000 passengers with 100 tonnes of cargo each) would then embark on the 40-400 (average of 150) day transit to Mars. The 1,000 Starships would land on the surface of Mars for unloading and then loading for the return trip to Earth. For the return trip, an extended time period also applies [McFall-Johnson, 2020]. Therefore, for a complete round trip to Mars and back would take approximately five years. At this pace, it would take 50 years to transport one million people to Mars. However, a smaller goal of only 100,000 would take only 5 years to transport the people to Mars. [Musk, 2017] Currently, each individual SpaceX Mars mission transport spacecraft would include: logistics support launches such as: one passenger rocket with mission Starship (100 people), three fuel rockets, and one cargo rocket. One Mars Starship mission equates to five operational rocket launches from Earth.

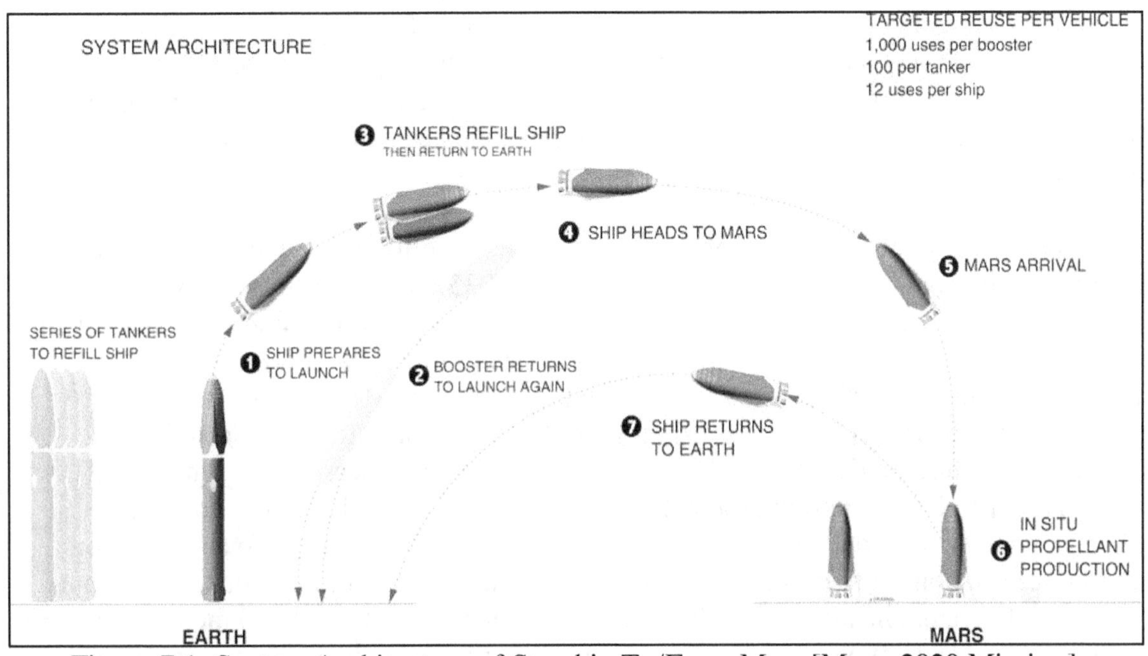

Figure 7.1: System Architecture of Starship To/From Mars [Mars, 2020 Mission]

The new few points illustrate the operational concept of using Space X's Starship (Notes from SpaceX's website):

1)   A series of tankers would be ready to refill the mission Starship. They would be launched into LEO to refill the Starship and then return to Earth for refilling. It is envisioned that the propellant tanker would go up anywhere from three to five times to fill the tanks of the Starship in orbit. In addition, there is the logistics supply ship to supply for the trip to Mars.

2)   Mission Starships would launch from prescribed launching pads on Earth with payloads (people, materials) and would be fueled in LEO by the tankers.

3)   Once fueled, there would be 1,000 Starships making the journey to Mars every 26-months. This way the Mars fleet would depart in masse to Mars over an 8 week period (or an average of 18 trans-Mars injections per day.

4)   Once arriving at the destination, the rocket would land on the surface to unload.

5)   Once refurbished, refueled, and reloaded, the rocket would make the return trip to Earth. (NOTE: Propellant would be loaded on Mars for the return trip after being produced from the atmosphere and in situ resources).

6)   The rocket would land on the Earth to unload and be refurbished.

7)   Repeat steps 1 – 6 as required.

To simplify these numbers, if the number is one million people with one ton per person, the common factor would be a mission Starship's capability to deliver to Mars 100 people and 100 tonnes. If those assumptions are a good starting point, then it would take 10,000 mission Starships to deliver the stated quantity of payload and passengers. As explained, the need for five launches for each mission Starship to depart for Mars, that would be approximately 50,000 launches. Mr. Musk has also stated his objective to be able to launch three times a day. Without factoring in the clustering necessary to match launch windows with Mars, the total number of years for those 50,000 launches would be: 16,667 days or 45.6 years.

7.5.2   Galactic Harbour Movement of Tonnage to Mars: With the understanding that the initial Space Elevators will not be available until after 2035, the flow of mass towards Mars is launch dominated early on with a transition to the permanent infrastructure as it becomes available. The vision is six mature Space Elevators inside three Galactic Harbours with each Space Elevator providing 79 tonnes per day towards Mars. This reflects the maturation from one Initial Operational Capability Space Elevator to three Galactic Harbours with mature Space Elevators (see Figure 7.2). When assessing the demands of customers for delivery to mission orbits, the strengths of the Space Elevator show their dominant characteristic - delivery of massive cargo to orbit. In the study "Space Elevators are the Transportation Story of the 21st Century," calculations showed the total throughput for six Space Elevators inside three Galactic Harbours ends up with 170,000 metric tonnes delivered to GEO and beyond each year. To supply one million tonnes, the mature Galactic Harbour would take about six years to fulfill the needs.

7.6 Comparison of Dual Space Access Architecture:

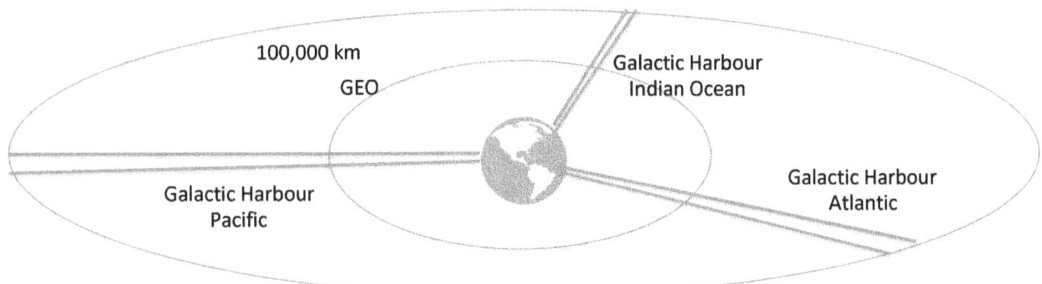

Figure 7.2, Vision of Three Galactic Harbours

- SpaceX alone supplies Mars: As shown previously, One million tonnes and one million people would take 50,000 launches and 45 years. (as an estimate)

- Galactic Harbour alone supplies Mars: For comparison's sake, the use of only Space Elevators is calculated. The bottom line is that the delivery of one million tonnes to Mars would take approximately 6 years -- (1,000,000/79 per day) / 6 SEs = 2110 days/365 or 5.7 years. As this report has shown in previous chapters, the Space Elevator is carbon negative so it will have a "beneficial effect" by reducing the 50,000 launches number to a fraction of 50,000.

- Dual Space Access Architecture (DSAA) Supplies Mars: As shown in the comparison of the two individual approaches just above, the Dual Space Access Architecture could break out sharing of the load such as 75% of cargo by the Space Elevators with humans and 25% of the cargo to go by future rockets. This will avoid the majority of the rocket equation relating to 75% of mass delivery and also allows the Starships to deliver principally critical cargo, such as people. This could be accomplished once the colony is established and Space Elevators are operational. If we eliminated 75 percent of the Starship launches and focused upon people movement with the other 25 percent of estimated launches it would show: a quadrupling of people to 400 per Starship. This would result in only 2,500 mission Starships and an additional 10,000 launches in support. These 12,500 launches, following the DSAA approach, would be saving 37,500 launches from the "only Starship approach." In parallel, the Space Elevators would be delivering the one million cargo tonnes to Mars in only six years (or 3/4ths of the delivered mass without passengers). The remarkable result is that there are 37,500 fewer launches along with 9,495 Space Elevator lift-offs. The next figure showing a comparison of the multiple factors influencing the use of Space Elevators and future rockets shows a remarkable story. The best answer is to move from rockets only to a Dual Space Access Architecture.

Approach	StarShips Only	Galactic Harbour Only	Dual Space Access Architecture - Rockets	Dual Space Access Architecture - Space Elevators
People	1,000,000	0	1000000	0
Mass (tonnes)	1,000,000	1,000,000	250,000	750,000
Amount per trip (people / tonnes]	100 / 100	0 / 79	400	0 / 79
Number of S/C to Mars	10,000	12,660	2500	9495
# Launches per mission S/C	5	1	5	1
Number Launches	50,000	12,660	12,500	9495
Launches per Day	3	6	3	6
Years to Accomplish	45.7	5.7	11.4	4.4

Table 7.4, Comparison of Liftoffs and Launches

In addition, there are so many other factors that are improved by using a permanent transportation infrastructure; such as,

- Departure Schedule: As per the discussion in the ISEC study report "Space Elevators are the Transportation Story of the 21st Century," the Arizona State University research conclusions showed that Space Elevators allow for daily departures for Mars. (no 26 month wait) [Swan, 2019]
- In addition, the ability of the Space Elevator to provide tremendous velocity (7.76 km/sec) upon release from the Apex Anchor, the delivery time varies over the year, but can be a short as 61 days. The average is somewhere in the 120-150 day range. Fast transit of cargo can be perceived as a train schedule.
- Although the cost is coming down for reusable launch vehicles, the cost of delivery by Space Elevators is revolutionary low.
- The environmental impacts of Space Elevators are carbon negative and very Earth Friendly with Space Elevators.
- The Galactic Harbour infrastructure, with six Space Elevators, will be available any day of the week around the equator, ready to liftoff for mission destinations.
- Value delivered can be compared to FedEx, Amazon, and DHL - they can deliver rapidly, on time, or as you would like it.
- Safety and reliability are watchwords for permanent transportation infrastructure, think bridges and train tracks.
- The exploitation of resources is very low, as the tether climbers use the solar energy to power the wheels and do not use consumables. The efficiency of both the cost factor and the resource factor is remarkably high, especially when compared to a rocket equation approach where the consumption of everything is dominant.

Comparison of Journeys: When embarking on a long journey, each of us thinks about the destination first. However, to have a successful journey, one must also consider the other factors that will be impacted; such items as: cost, travel time, environmental impact, availability, value delivered, safely and resource efficiency. The previous table shows the relationships between the current rockets, future rockets and Space Elevators.

> Concept: When Space Elevators are ready, the factors influencing
> movement of mission payloads will not be dominated by resource
> consumption and environmental impacts.

One message from this analysis is that if there are thousands of launches per year to support our future missions to Mars alone, then the impact on the environment must be evaluated. In addition, just by the reality of the rocket equation, that approach consumes precious resources in great quantities. The example of the Apollo mission is illuminating in complexity of missions leveraging rockets against the Earth's gravity well. The study authors concluded that, the future will lead to rockets leveraged for special cargo, special orbits and movement of people. Space Elevators will do all the heavy lifting and routine delivery of cargo.

7.7 Realization:

A Dual Space Access Architecture, combining rocket and Space Elevator strengths, results in tremendous advantages in the "greening of the Earth." The first is that all the robotic movement of mass (cargo, habitats, air, water, etc) would be moved safely, routinely, daily, environmentally friendly, and inexpensively by Space Elevators. The second strength of DSAA is this separation of delivery approaches which greatly enhances missions in the future. As customer demands for huge masses matures to support near term missions such as Space Solar Power (five million tonnes to GEO) and a Mars Colony (one million tonnes to Mars), the value of Space Elevators becomes obvious. When the Space Elevator delivers 75% of the mass needed for critical missions, the savings in cost, time and environmental impact will make us ask: Why not sooner?

> Space Elevators and Galactic Harbours are Big Green Machines designed to improve
> the Earth's environment through two significant contributions. The first is the
> remarkable "zero-emission" lift of cargo to space - reducing environmental impacts
> from rocket launches. The second is the ability to deploy massive systems to GEO
> and beyond that eliminate rocket launches by becoming a partner in Dual Space
> Access Architecture.

7.8 Conclusions:

The net assessment trade study conducted by ISEC showed that the conclusions from this chapter are threefold:

Conclusion #1:   Space Elevator Vision matches the aggressive visions and missions "out there" for movement off planet and improving the Earth's environment. We are building the Green Road to Space in response to some of the current vision. We want to be the second lane in a dual lane road to space.

> Blue Origin's Vision: "Millions of people living and working in space"
>   and "I am going to build the road to space.
> "SpaceX's Vision: "Making Humanity Multi-planetary"
> National Space Society Vision; "People living and working in thriving communities
>   beyond the Earth, and the use of the vast resources of space
>   for the dramatic betterment of humanity."
> "ESA Director General Jan Wörner and NASA Administrator Jim Bridenstine have
> signed a Memorandum of Understanding (MoU) to take Europe to the Moon." (ESA
> Announcement 28 Oct 2020).

Now, what does the new movement off-planet do to/for us as Space Elevator enthusiasts? It reinforces our critical nature as participants in the future. If everyone wants to have their citizens living on the Moon (and Mars of course), massive movement of equipment needs to occur. Space Elevators are the answer! Space Elevator teams need to have a vision, inside the Space Elevator community, that is supportive of this historic achievement and bring us into their embryonic endeavors.

Conclusion #2:   The approach to development of a Dual Space Access Architecture must be initiated.

With the concept of Galactic Harbours comes the recognition that movement off-planet will require complementary capabilities, such as rockets portals and Galactic Harbour infrastructures, each with their own strengths and short-falls.   While editing the latest ISEC year-long study report, the authors recognized some powerful truths. One of the biggest is that Space Elevators will stand up strong next to rockets and help enable movement off-planet. When the authors look at the Moon and dream of spaceflight, they forget how extremely difficult it was to accomplish - both in energy and design complexity. By having a joint approach to movement beyond Low Earth Orbit, the hopes, visions, and dreams can come true.

Conclusion #3:   Recognize the potential benefits of Space Elevators when reaching for the stars.

- Potential Beneficial Impact of Space Elevator: Reducing the number of launches (such as to support humanity's movement off planet) will decrease pollution significantly while avoiding the rocket equation.
- Potential Beneficial Impact of Space Elevator: Provides safe, reliable, routine, daily, environmentally friendly, and inexpensive transportation infrastructure to move massive tonnage to GEO and beyond, specifically the Moon and Mars.

7.9 Recommendation:

Conduct global and future launch flotilla studies inside a Dual Space Access Architecture framework. As shown in the conclusion, there will be hundreds, or even thousands, of rocket launches in the not to distant future to support the dreams and plans for Space-Based Solar Power and Missions to the Moon and Mars. This increase from roughly 100 launches per year around the globe to thousands from one or two launch sites could be sooner than we think. Therefore; the acceptance of the Space Elevator must be sooner rather than later.

7.10 Summation:

Fifty plus years after Apollo, the human race has decided to create a permanent presence on the Moon, in space, and on Mars. I suggest our vision should be something like the original statement in this paper and shown to include three Galactic Harbours, with two Space Elevators each, leading to 170,000 tonnes to GEO and beyond.

> Space Elevators are the Green Road to Space as they enable humanity's most important missions by moving massive tonnage to GEO and beyond. They accomplish this safely, routinely, inexpensively, and daily; they are environmentally neutral.

In addition, we must dream big and see the Space Elevator of the future. As we discussed in our last study report, ISEC feels:

> "The Space Elevator story is still being written. The Apex Anchor is where the Space Elevator meets the Shoreline of Outer-Space and Where the Transportation Story of the 21st Century meets the Final Frontier."

Chapter 8 - Conclusions

8.1 Introduction:

The question in the space arena should not be - how do we build bigger and better rockets to support these customer demands? - because the massive movement of cargo can never become efficient for rockets. One must move to the Dual Space Access Architecture concept with Space Elevators moving massive tonnage while the Galactic Harbour encourages and develops space enterprises along their vertical "train tracks." Rockets, as complementary to Space Elevators, have their place when delivering payloads to LEO or MEO. It is basically economical with minimal impact to the Earth's environment. However, when venturing beyond LEO or MEO, to GEO and other planets it simply is not feasible to "build a bigger" rocket that has to make tens of thousands of launches to deliver the required payloads. Utilizing the capabilities of Space Elevators, coupled with rockets to create a "Dual Space Access Architecture" is the most efficient, cost effective way to deliver payloads outside Earth's neighborhood.

8.2 Four Study Conclusions:

The net assessment trade study conducted by ISEC showed that the basic conclusions from this report are three fold:

Conclusion #1: The Space Elevator Vision matches the aggressive visions and missions "out there" for movement off planet and improving the Earth's environment. We are building the Green Road to Space in response to some of those current visions. Elon Musk to Mars! Dr. Mankins eliminating coal planets. Colonies on the Moon. We want to join those visionaries and become the second lane in a dual lane road to space. One example from Blue Origin is:

> "Millions of people living and working in space"
> and "I am going to build the road to space."

Now, what does the new movement off-planet do to/for us as Space Elevator enthusiasts? It reinforces our critical nature as participants in the future. If everyone wants to have their citizens living on the Moon (and Mars of course), massive movement of equipment needs to occur. Space Elevators are the answer! Space Elevator teams need to have a vision, inside the Space Elevator community, that is supportive of this historic achievement and bring us into their embryonic endeavors. We CAN move millions of tonnes of cargo – no one else can with an environmental approach with timely delivery.

Conclusion #2: The approach to development of a Dual Space Access Architecture must be initiated.

With the concept of Galactic Harbours comes the recognition that movement off-planet will require complementary capabilities, such as rockets portals and Galactic Harbour infrastructures, each with their own strengths and short-falls. While editing the latest ISEC year-long study report, the authors recognized some powerful truths. One of the biggest is that Space Elevators will stand up strong next to rockets and help enable movement off-planet. When the authors look at the Moon and dream of spaceflight, they forget how extremely difficult it was to accomplish - both in energy and design complexity. By having a joint approach to movement beyond Low Earth Orbit, the hopes, visions, and dreams can come true.

Conclusion #3: Recognize the potential benefits of Space Elevators when reaching for the stars.

Table 8.1:   Potential Beneficial Impacts of Space Elevator:

Approach	Effects
Enabling Space Solar Power	Reducing the number of fossil fuel burning plants providing energy (100s of coal plants) by using the delivery of energy from orbit to anywhere all the time
Zero (or negative) carbon footprint to deliver to space	Daily operations, at zero (or negative) carbon footprint, reduces the environmental impact of the expected massive movement to space.
Enable Appropriate Solar Shade at L-1	Reducing the energy from the Sun that reaches the Earth's Atmosphere, thus reducing global warming.
Reduce Burning of fuel in Atmosphere	Replacing number of rocket launches (such as to support humanity's movement off planet) will decrease pollution significantly.
Environmentally Friendly Space Infrastructure	Provides safe, reliable, routine, daily, environmentally friendly, and inexpensive transportation infrastructure to move massive tonnage to GEO and beyond, specifically the Moon and Mars.
Enable Permanent Disposal of High Level Nuclear Waste	Deposits Nuclear Waste in Small Solar Orbit which provides safe and long term storage of High-Level Nuclear Waste.

Conclusion #4: While looking at the topic of environmental impact of Dual Apace Access Architecture, the study team recognized that there was no consolidated evaluation of rocket launch impact on the global environment. With the future offering thousands of launches per year and three a day at some locations an additional question must be: What are the environmental effects of thousands of launches per year upon topics such as burning rocket fuel in the atmosphere, residual space debris from each launch, and black carbon impact on the ozone layer. The increase from one hundred launches per year to thousands deserves a major study looking at their global impacts.

8.3 Realization: A Dual Space Access Architecture Leads to "greening of the Earth:" The first is an insight, combining rocket and space elevator strengths results in tremendous advantages in that: rockets are here now and robust resulting in rapid transit through radiation belts with people, and second is that all the robotic movement of mass in the future (cargo, habitats, air, water, etc) would be moved safely, routinely, daily, environmentally friendly, and inexpensively by Space Elevators. This separation of delivery approaches will greatly enhance missions. As customer demands for huge masses matures to support near term missions such as Space Based Solar Power (five million tonnes to GEO) and a Mars Colony (one million tonnes to Mars), the value of Space Elevators becomes obvious. When the Space Elevator delivers 75% of the mass needed for critical missions, the savings in cost, time and environmental impact will make us ask: Why not sooner?

This study will start the future discussions by showing the additional benefits of Space Elevators being defined as "Massive Green Machines." The current vision is:

> " Space Elevators are the Green Road to Space while they enable humanity's most important missions by moving massive tonnage to GEO and beyond."

In point of fact, the operations of Space Elevators and Galactic Harbours will be carbon negative. Several of the concepts developed above can be considered key elements establishing the reality that Space Elevators can make the Earth Greener.

This net assessment trade study conducted by ISEC showed that:

> Space Elevators and Galactic Harbours are Big Green Machines designed to improve the Earth's environment through two significant contributions. The first is the remarkable "zero-emission" lift of cargo to space - reducing environmental impacts from rocket launches. The second is the ability to deploy massive systems to GEO and beyond that eliminate rocket launches by becoming a partner in Dual Space Access Architecture.

Chapter 9 Recommendations

Implement Dual Space Access Architecture and Space Elevators as Big Green Machines

Rockets to open up the Moon and Mars with Space Elevators to supply and grow the colonies. In addition, Space Elevators will enable Green Missions such as, Space Solar Power and L-1 Sun Shade. This compatible and complementary approach with future rockets is not competitive while leveraging the strengths of both inside a Dual Space Access Architecture.

Indeed, the authors see others ready to leap into the off-planet movement Space Elevators should offer their zero carbon footprint approach. Space Elevators have tremendous strengths that have not yet been included in others' strategies for going to the Moon and beyond. This new movement off planet should include Space Elevators' ability to:

- Depart the Apex Anchor at great velocity (7.76 km/sec)
- Support interplanetary missions (Fast Transit to Mars 61 days)
- Supply massive daily payloads (170,000 tonnes per year)
- Create entrepreneurial enterprises along the Galactic Harbour
- Enable new environmentally significant missions (Space Solar Power, Solar Shades, hi-level nuclear waste disposal, etc.)
- Enable carbon negative operations for delivery to orbit
- Exit the gravity well and avoid the burden of the rocket equation
- And, accomplish this daily, routinely, inexpensively and carbon negatively.

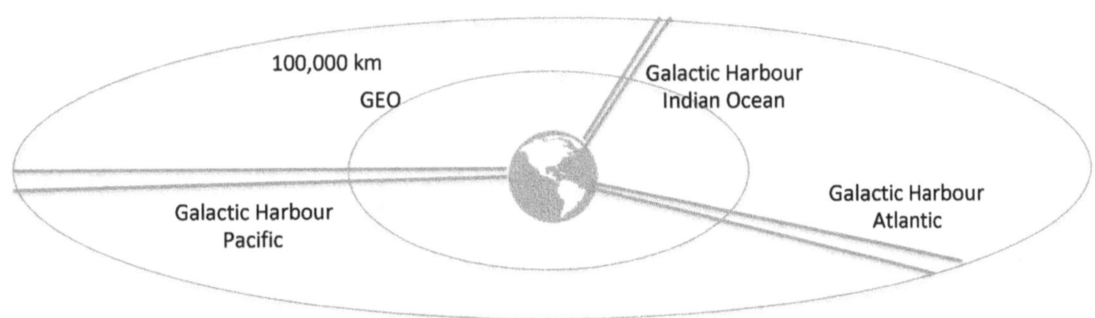

In addition, the team must dream big and see the Space Elevators of the future. Our last study report expressed its future as:

"The Space Elevator story is still being written. The Apex Anchor is where the Space Elevator meets the Shoreline of Outer-Space and Where the Transportation Story of the 21st Century meets the Final Frontier."

References

[3%] World Nuclear Association https://www.world-nuclear.org/nuclear-essentials/what-is- nuclear-waste-and-what-do-we-do-with-it.aspx 18 02 2020

[13,865 in 2019] STOCKHOLM INTERNATIONAL PEACE RESEARCH INSTITUTE, 17 June, 2019 https://www.sipri.org/media/press-release/2019/modernization-world-nuclear-forces- continues-despite-overall-decrease-number-warheads-new-sipri 21 08 2020

[Agreement, 1989] https://www.hanford.gov/page.cfm/TriParty/TheAgreement and https://www.hanford.gov/page.cfm/TriParty/TheAgreement 11 21 2020 (Note: Includes amendments through Sept. 30, 2019)

[Angel, 2006a] R. Angel, S.Pete Worden, "Making Sun-Shades from Moondust - Global warming and NASA's new vision could share a common solution far above the Earth," From Ad Astra, Volume 18 Number 1, Summer 2006

[Angel, 2006b]   Roger Angel, Feasibility of cooling the Earth with a cloud of small spacecraft near the inner Lagrange Point (L1) PNAS, Nov 14, 2006) Vol. 103, No 46.

[anti-nukes, 2018] Reuters World News April 18, 2018 https://uk.reuters.com/article/uk-france- nuclearpower-waste/quiet-no-more-french-village-becomes-centre-of-anti-nuclear-protest-idUKKBN1HP1S7 06 12 2020

[Apollo, 1969] Description of equipment for Apollo 11 trip, https://en.wikipedia.org/wiki/Apollo_11

[Artsutanov, 1960] Artsutanov, Y. V Kosmos na Electrovoze, Komsomolskaya Pravda, July 31, 1960.

[Bewicke, 2019] Bewicke, Henry, "What is the Environmental Impact of SpaceX?" 2002 Dec19

[Biden, 2020] Joseph R. Biden and Bernard Sanders. Biden-Sanders Unity Task Force Recommendations, 2020.

[Bouchard, 2019] Bouchard, Anthony, "How Rocket Launches Impact the Environment", Labroots, 18 Jun 2019

[Burns, 1978] 1978NASA "Nuclear waste disposal in space", Burns, R.E. et al , https://ntrs.nasa.gov/citations/19780015628

[Cannon, 2019] Cannon, Kevin; Britt, Daniel, "Feeding One Million People on Mars," Space News, 20Aug19, https://doi.org/10.1089/space.2019.0018

[Carter, 1997] Rossin, Dr. A. David, U.S. Policy on Spent Fuel Reprocessing 1977 https://www.pbs.org/wgbh/pages/frontline/shows/reaction/readings/rossin.html 11 08 2020

[Cash, 2017] I. Cash, "CASSIOPeiA solar power satellite," in 2017 IEEE International Conference on Wireless for Space and Extreme Environments (WiSEE), (Montreal, QC), pp. 144–149, IEEE, Oct. 2017.

[Casks 2020] https://www.world-nuclear.org/information-library/country-profiles/countries-t-z/usa-nuclear-fuel-cycle.aspx 03 11 2020

[Cha, 2020] S. Cha, "South Korea's Moon targets carbon neutrality by 2050." Reuters, Oct 2020.

[China, 2020] "China aims to cut its net carbon-dioxide emissions to zero by 2060," The Economist, Sept. 2020.

[consensus] World Nuclear Association https://world-nuclear.org/information-library/nuclear- fuel-cycle/nuclear-wastes/storage-and-disposal-of-radioactive-wastes.aspx 14 03 2020

[Cordes, 2016] Cordes, Erik, et.al., Environmental Impacts of Deep-Water Oil and Gas Industry: A Review Guide Management Strategies," Frontiers in Environmental Science, 16Sep16

[crude oil 2020] World Nuclear Association https://www.world-nuclear.org/information-library/facts- and-figures/heat-values-of-various-fuels.aspx 02 10 2020

[David, 2020] L. David and M. N. Ross, "An Underappreciated Danger of the New Space Age: Global Air Pollution, "Scientific American, Nov. 2020.

[Denyer, 2020] Simon Denyer and Akiko Kashiwagi. Japan, world's third largest economy, vows to become carbon neutral by 2050. Washington Post, October 2020.

[DGR, 2020] https://world-nuclear.org/information-library/nuclear-fuel- cycle/nuclear-wastes/storage-and-disposal-of-radioactive- wastes.aspx 03 14 2020

[DGR stalled 2018] Eye On The Arctic, 23 01 2018 https://www.rcinet.ca/eye-on-the-arctic/2018/01/23/environmentalists-praise-ruling-nuclear-waste-site/ 26 09 2020

[Dry] Radioactivity.Eu.com Storing: a useful but temporary solution https://www.radioactivity.eu.com/site/pages/Temporary_Storages.htm 02 12 2020

[Edwards, 2003] Bradley C Edwards and Eric A. Westling, The Space Elevator: A Revolutionary Earth-to-Space Transportation System. BC Edwards, 2003.

[EU, 2019] "EU carbon neutrality: Leaders agree 2050 target without Poland" BBC News, Dec. 2019

[fiasco 2013] DARIUS DIXON, Politico Magazine 11/30/2013 https://www.politico.com/story/2013/11/nuclear-waste-fiasco-100450 11 04 2020 John Sadler, Las Vegas Sun May 19, 2020

[Finland, 2020] https://www.nsenergybusiness.com/projects/onkalo-nuclear- waste-disposal-facility/ 05 29 2020

[Finland, DGR] NS ENERGY https://www.nsenergybusiness.com/projects/onkalo-nuclear-waste- disposal-facility/ 06 09 2020

[fission products] https://www.radioactivity.eu.com/site/pages/Fission_Products.htm 19 03 2020

[Fitzgerald, 2015] Fitzgerald, Michael; Penny, Robert; Swan, Peter; Swan, Cathy; "Space Elevator Architecture and Roadmaps: A Primer for Progress in Space Elevator Development", ISEC Position Paper # 2014-1, 2015

[Fitzgerald, 2017] Fitzgerald, Michael; Hall, Vern; Swan, Pater; Swan, Cathy; "Design Considerations for the Space Elevator Apex Anchor and GEO Node", ISEC Position Paper #2017-1

[Fitzgerald, 2019] Fitzgerald, M and V. Hall Portions of this chapter leveraged vision statement, Dec 2019 (see appendix).

[Folga, 1996] Folga, S. M. et al, High-Level Waste Inventory . . . Dec. 1996 30001802.pdf ppg. 92-99

[French, DGR] World Nuclear Association Sept. 2020 https://www.world-nuclear.org/information-library/country-profiles/countries-a-f/ france.aspx (Radioactive waste management), 06 12 2020

[Fusion, 2020] https://www.iter.org/sci/Fusion 06 06 2020

[Gerber, 1992] 10144167.pdf M. S. Gerber "Legend and Legacy: Fifty Years of Defense Production at the Hanford Site" 1992, ppg. 22, 27, 32 - 37

[Gen IV] AMERICAN NUCLEAR SOCIETY Position Statement 74 November 2005 ps74.pdf p 1; International Atomic Eneregy Agency, No. NW-T-1.7 Waste from Innovative Types of Reactors and Fuel Cycles PUB1822_web.pdf ppg. 3-10; World Nuclear Association https://world-nuclear.org/information-library/nuclear- fuel-cycle/nuclear-power-reactors/generation-iv-nuclear-reactors.aspx 03 06 2020 Danilo Nori INFN - Istituto

Nazionale di Fisica Nucleare IV Generation Nuclear Reactors,
IV_Generation_Nuclear_Reactors.pdf p 1-5; University of Calgary, Energy Education
https://energyeducation.ca/encyclopedia/Generation_IV_nuclear_reactors#:~:text=T
he%20Gen-IV%20reactors%20are%20being%20researched%20and20developed,
South%20Korea%2C%20South%20Africa%2C%20Switzerland%2C
%20and%20the%20EU. 26 11 2020 Columbia University in the City of New York GEN I
V: Evolution in Nuclear Safety July 15, 2012 https://k1project.columbia.edu/a15
Gen IV Nuclear Energy Systems Services, LLC
https://www.genivnuclearenergysystems.com/home.html 16 11 2020 World Nuclear
Association https://www.world-nuclear.org/information- library/country-profiles/countries-
o-s/russia-nuclear-power.aspx 15 06 2020

[Glaser, 1968] Peter E. Glaser. Power from the Sun: Its Future. Science,
162(3856):857–861, November 1968.

[Hanley, 1978] G. Hanley, "Satellite Power Systems (SPS) Concept Definition Study,"
tech. rep., Apr. 1978.

[iaea; 60%] International Atomic Energy Agency
https://www.iaea.org/newscenter/multimedia/videos/iaea-data-animation- meeting-climate-
goals-as-more-countries-embark-on-nuclear-power 09 05 2020

[iaea, Dumping] International Atomic Energy Agency, Bulletin 4/1 1989 31404684750.pdf,
ppg. 47-50

[iea, 2018] IEA https://www.iea.org/data-and-statistics/charts/world-gross-electricity-
production-by-source-2018 26 11 2020

 [large casks] World Nuclear Association, Sept. 2020 https://www.world-
nuclear.org/information-library/country-profiles/countries-t-z/usa- nuclear-fuel-cycle.aspx
11 03 2020

[Levykin, 2018] In 2018, rocket engines emitted 10,000 tons of black carbon and
aluminium particles into the stratosphere, about the same annual amount caused by global
aviation (9,500 tons). Levykin, Volodymyr, "Going Green: Why the Launch Industry
Urgently Needs Environmental Regulations," ViaSatellite Podcast, Aug 19, 2020.

[low cost] World Nuclear Association https://www.world-nuclear.org/information-
library/economic-aspects/economics- of-nuclear-power.aspx 18 09 2020

[Mankins, 2002] John C Mankins. A TECHNICAL OVERVIEW OF THE "SUNTOWER"
SOLAR POWER SATELLITE CONCEPT. Acta Astronautica, 50(6):369–377, March 2002.

[Mankins, 2006] Mankins, John Dr., "Broadcast 609 on 17Oct06 or "The Case for Space
Solar Power", Virginia Publishing)

[Mankins, 2011] Mankins, John; Kaya, Nobuyuki (eds.) (2011). The First International Assessment Of Space Solar Power : Opportunities, Issues And Potential Pathways Forward. International Academy of Astronautics. p. 249.

[Mankins, 2012] "Solar space power: the tipping point:" John C. Mankins at TEDxGood enoughCollege.

[Mankins, 2012] J. Mankins, N. Kaya, and M. Vasile, "SPS-ALPHA: The First Practical Solar Power Satellite via Arbitrarily Large Phased Array (A 2011-2012 NIAC Project)," in 10th International Energy Conversion Engineering Conference, (Atlanta, Georgia), American Institute of Aeronautics and Astronautics, July 2012.

[Mankins, 2013] John C Mankins. The Case for Space Solar Power. Virginia Edition Publishing, LLC, 2013.

[Mankins, 2017] John C Mankins. New Developments in Space Solar Power. NSS Space Settlement, (3), December 2017.

[Mankins, 2019] Mankins, John, Conversation with Dr. Peter Swan at the International Astronautical Congress (in Washington DC), 2019.

[Mankins, 2020] Mankins, John, Webinar entitled "NSS Space Forum - A Case for Space-Based Solar Power," 20 August 2020.

[Mars 2020 Mission] "Environmental Consequences," Draft Environmental Impact Statement for the Mars 2020 Mission, 2019

[McFall-Johnson 2020] McFall-Johnson, Morgan, Dave Mosher, "Starship rockets every day and creating a log of jobs on the red planet," Business Insider, 17Jan 20.

[Miraux, 2020] Miraux, Lois, "Why addressing the environmental crisis should be the Space industry's Top Priority," The Space Review, Oct 5, 2020, web source.

[Musk 2017] Musk, Elon, "Making Humans a Multi-Planetary Species", New Space, Vol 2, No. 2, 2017

[Nathaniel 2017] Nathaniel, "The Ugly Side of Space Exploration," The Startup, 16 June 2017.

[near zero] World Nuclear Association https://www.world-nuclear.org/information-library/energy-and-the- environment/co2-implications-of-electricity-generation.aspx 04 10 2020

[Newcomb, 2019] Newcomb, Tim 10 12 2019 https://www.popularmechanics.com/science/energy/a30184557/hanford-nuclear-waste-site-cleanup/ 11 21 2020

[NREL, 2020] "City Energy Data APIs | NREL: Developer Network," Nov. 2020. https://developer.nrel.gov/docs/cleap/.

[one-fifth] U.S. NUCLEAR WASTE TECHNICAL REVIEW BOARD VITRIFIED HIGH-LEVEL RADIOACTIVE WASTE vitrified_hlw.pdf; (ppg. 1 - 4)

[O'Neill, 1974] O'Neill, Gerard K. (September 1974). "The Colonization of Space". Physics Today. 27 (9): 32–40. Bibcode:1974PhT....27i..32O. doi:10.1063/1.3128863.

[Pachauri, 2015] R.K. Pachauri, Leo Nayer and Intergovernmental Panel on Climate Change, editors. Climate change 2014: synthesis report. Intergovernmental Panel on Climate Change. Geneva, Switzerland, 2015. OCLC: 914851124
change 2014: synthesis report. Intergovernmental Panel on Climate Change, Geneva, Switzerland,

[peak 1986] Our World in Data https://ourworldindata.org/nuclear-weapons 04 28 2020

[Peet, 2021] Peet, Matthew, "The Orbital Mechanics of Space Elevator Launch Systems," Acta Astronautica, Vol 179, Feb 2021, pg 153-171. [Penny 2012] Penny, Robert; Peter Swan, Cathy Swan;"Space Elevator Concept of Operations," ISEC Position Paper # 2012-1

[Penny, 2015] Penny, Robert; Hall, Vern; Glaskowsky, Peter; Schaeffer, Sandee; "Design Consideration of a Space Elevator Earth Port: A Primer for Progress in Space Elevator Development," ISEC Position Paper # 2015-1

[Pearson, 1975] Jerome Pearson. The orbital tower: A spacecraft launcher using the Earth's rotational energy. Acta Astronautica, 2(9-10):785–799, September 1975.

[Pearson, 1997] Jerome Pearson. Konstantin Tsiolkovski and the Origin of the Space Elevator. volume 48, Turin, Italy, October 1997.

[Population, 2020] Population total, World Bank Data Portal, July 2020

[Proliferation] Erin Blakemore, 05 11 2020 How the advent of nuclear weapons changed the course of history https://www.nationalgeographic.co.uk/history-and-civilisation/2020/07/how-the- advent-of-nuclear-weapons-changed-the-course-of-history 25 11 2020 Hoover Institution The History Of Nuclear Warfare And The Future Of Nuclear Energy, Friday, March 15, 2019 https://www.hoover.org/research/history-nuclear-warfare-and-future-nuclear-energy 25 11 2020 The Editors, History.com, Cold War History https://www.history.com/topics/cold-war/cold-war-history?li_source=LI&li_medium=m2m-rcw-history 14 06 2020

[Rastogi 2009] Rastogi, Nina Shen "Dirty Rockets: What's the Emvironmental Impacts of Going into Space?," Tweet & Slate Website, 17Nov09

[Reprocess] Radioactivity.Eu.com Fuel reprocessing
https://www.radioactivity.eu.com/site/pages/Reprocess_Spent_Fuel.htm 14 05 2020

[Repackage] Radioactivit,.Eu.com Waste Conditioning
https://www.radioactivity.eu.com/site/pages/Waste_Conditioning.htm 26 10 2020

[Siegel 2019] Siegel, Ethan, Senior Contributor Group Science, "Starts With A Bang" 20 09 2019 https://www.universetoday.com/133317/can-we-launch-nuclear-waste- into-the-sun/ 11 03 2020

[Sasaki, 2007] S. Sasaki, K. Tanaka, K. Higuchi, N. Okuizumi, S. Kawasaki, N. Shinohara, K. Senda, and K. Ishimura, "A new concept of solar power satellite: Tethered-SPS," Acta Astronautica, vol. 60, pp. 153–165, Feb. 2007.

[Sittlow, 2020] Sittlow, Brian L. Council on Foreigh Relations, New START: The Future of U.S.- Russia Nuclear Arms Control https://www.cfr.org/in-brief/new-start-future-us-russia-nuclear-arms-control 11 07 2020

[sole repository] YuccaMountain.org, https://yuccamountain.org/time.htm 11 04 2020

[SpaceX, 2020] "SpaceX Starship Users Guide, Rev 1.0" SpaceX website, Mar 2020.

[Swan, 2010] Swan, P. "First Space Elevator: on the Moon, Mars or the Earth?" IAC-10-D4.4.10, 2010.

[Swan, 2020c] Swan, P. "Earth Space Elevator First," Space Elevator Newsletter, 2020.

[Swan, 2013] Peter A Swan, David I Raitt, Cathy W Swan, Robert E Penny, John M Knapman, and Acad´emie Internationale d'Astronautique. Space elevators: an assessment of the technological feasibility and the way forward. International Academy of Astronautics Paris, 2013

[Swan, 2014] Swan, P., Raitt, Swan, Penny, Knapman. International Academy of Astronautics Study Report, Space Elevators: An Assessment of the Technological Feasibility and the Way Forward, Virginia Edition Publishing Company, Science Deck (2013) ISBN-13: 978-2917761311

[Swan, 2019a] Swan, Peter; Fitzgerald, Michael, "Today's Space Elevator: Space Elevator Matures into the Galactic Harbour," ISEC Position Paper # 2019-1, Sep 2019, ISBN 978-0-359-93496-63

[Swan, 2019b] Swan, P., David Raitt, John Knapman, Akira Tsuchida, Michael Fitzgerald, Yoji Ishikawa, Road to the Space Elevator Era, Virginia Edition Publishing Company, Science Deck (2019) ISBN-19: 978-0-9913370-3-3

[Swan, 2017] Swan, Peter; Fitzgerald, Michael, "How the Space Elevator Grew into a Galactic Harbour", 68th IAC, 25-29 Sep 2017

[Swan, 2019a] Peter A Swan, David I Raitt, John M Knapman, Akira Tsuchida, Michael A Fitzgerald, and Yoji Ishikawa. Road to the Space Elevator Era. International Academy of Astronautics Paris, 2019.

[Swan, 2020b] Swan, Peter; Fitzgerald, Michael; Swan, Cathy, "Today's Space Elevator," Briefing, 30Apr20.

[Swan, 2020a] Swan, Pete, "Space Elevators are the Transportation Story of the 21st Century," ISEC Study Report, www.isec.org Mar20

[Swan, 2014] Swan, Peter; Swan, Cathy; Penny, Robert; Knapman, John; Glaskowsky, Peter, "Design Considerations for Space Elevator Tether Climbers: A Primer for Progress in Space elevator Development", ISEC Position Paper#2013-1, 2014, ISBN 978-1-312-09982-1.

[testing nukes] Arms Control Association "The Nuclear Testing Tally", July 2020 https://www.armscontrol.org/factsheets/nucleartesttally 07 08 2020

[towns compete] Tagliabue, John New York Times, July 9, 2018 Swedish towns vie for nuclear waste site https://www.sfgate.com/business/article/Swedish-towns-vie-for-nuclear-waste-site-3267892.php 08 09 2020

[UOA, 2006] "Space sunshade might be feasible in global warming emergency" Public Release 3 Nov 2006, Univ of Arizona - EurekAlert AAAS.

[USEIA, 2019] Office of Energy Analysis. International Energy Outlook 2019. Technical report, U.S. Department of Energy Information Administration, Sept. 2019. https://www.eia.gov/ieo.

[USSR] BBC News, Laurence Peter. Jan. 25, 2013 https://www.bbc.com/news/world-europe- 21119774 14 03 2020

[vitrified] U.S. NUCLEAR WASTE TECHNICAL REVIEW BOARD, "VITRIFIED HIGH-LEVEL RADIOACTIVE WASTE", Rev. 1, Nov., 2017 vitrified_hlw.pdf, ppg. 1-4

[warheads 2018] Stockholm International Peace Research Institute https://www.sipri.org/media/press-release/2019/modernization-world-nuclear- forces-continues-despite-overall-decrease-number-warheads-new-sipri 05 01 2020

[Wet] SKB https://www.skb.com/our-operations/CLAB/ 04 07 2020

[Whittaker, 2018] Whittaker, Ian, "Is SpaceX Being Environmentally REsponsible?" Smithsonian Magazine, 7 Feb 2018.

[WNA, 2020] World Nuclear Association https://www.world-nuclear.org/information-library/ facts-and-figures/reactor-database.aspx 07 05 2020 [Yucca, 1994] https://www.cbsnews.com/news/yucca-mountain-nuclear-waste- storage-controversy/ 06 14 2020

[WBDP, 2014] Electric power consumption (kWh per capita). World Bank Data Portal' n.d., 'Population, total. World Bank Data Portal' n.d.

[Yang, 2016] Yang Yang, Yiqun Zhang, Baoyan Duan, Dongxu Wang, and Xun Li. A novel design project for space solar power station (SSPS-OMEGA). Acta Astronautica, 121:51–58, April 2016.

[XU, 2020] Chi Xu, Timothy A. Kohler, Timothy M. Lenton, Jens-Christian Svenning, and Marten Scheffer. Future of the human climate niche. Proceedings of the National Academy of Sciences, May 2020. Publisher: National Academy of Sciences Section: Social Sciences.

[Zheleznogorsk, 2020] https://bellona.org/news/nuclear-issues/radioactive-waste-and-spent-nuclear-fuel/2015-07-hearings-on-russian-radwaste- repository-proceed-transparently-to-a-point 06 10 2020 & https://www.world-nuclear-news.org/WR-Russia-completes-design- of-underground-radwaste-research-laboratory-01041501.html 06 13 2020

Appendices

A - International Space Elevator Consortium
B - Avoiding the Rocket Equation,
C - Body of Knowledge
D - Studies (ISEC, IAA, Obayashi)
E - Hi-Level Nuclear Waste
F - Space Solar Power Backup

Appendix A - International Space Elevator Consortium

Who We Are
The International Space Elevator Consortium (ISEC) is composed of individuals and organizations from around the world who share a vision of humanity in space.

Our Vision
Space Elevators are the Green Road to Space while they enable humanity's most important missions by moving massive tonnage to GEO and beyond. This is accomplished safely, routinely, inexpensively, daily, and they are environmentally neutral.

Strategic Approach: Dual Space Access Architecture
Rockets to open up the Moon and Mars with Space Elevators to supply and grow the colonies. In addition, Space Elevators will enable Green Missions such as, Space Solar Power and L-1 Sun Shade. This compatible and complementary approach with future rockets is not competitive while leveraging the strengths of both inside a Dual Space Access Architecture.

Our Mission
The ISEC promotes the development, construction and operation of a space elevator infrastructure as a revolutionary and efficient way to space for all humanity.

What We Do
- Provide technical leadership promoting development, construction, and operation of space elevator infrastructures.
- Become the "go to" organization for all things space elevator.
- Energize and stimulate the public and the space community to support a space elevator for low cost access to space.

- Stimulate science, technology, engineering, and mathematics (STEM) educational activities while supporting educational gatherings, meetings, workshops, classes, and other similar events to carry out this mission.

A Brief History of ISEC

The idea for an organization like ISEC had been discussed for years, but it wasn't until the Space Elevator Conference in Redmond, Washington, in July of 2008, that things became serious. Interest and enthusiasm for a space elevator had reached an all-time peak and, with Space Elevator conferences upcoming in both Europe and Japan, it was felt that this was the time to formalize an international organization. An initial set of directors and officers were elected and they immediately began the difficult task of unifying the disparate efforts of space elevator supporters worldwide.

ISEC's first Strategic Plan was adopted in January of 2010 and it is now the driving force behind ISEC's efforts. This Strategic Plan calls for adopting a yearly theme to focus ISEC activities. Because of our common goals and hopes for the future of mankind off-planet, ISEC became an Affiliate of the National Space Society in August of 2013. In addition, ISEC works closely with the Japanese Space Elevator Association.

Our Approach

ISEC's activities are pushing the concept of space elevators forward. These cross all disciplines and encourage people from around the world to participate. The following activities are being accomplished in parallel:

- Yearly conference – International space elevator conferences were initiated by Dr. Brad Edwards in the Seattle area in 2002. Follow-on conferences were in Santa Fe (2003), Washington DC (2004), Albuquerque (2005/6 –smaller sessions), and Seattle (2008 to the present). Each of these conferences had multiple discussions across the whole arena of space elevators with remarkable concepts and presentations.
- Yearlong technical studies – ISEC sponsors research into a focused topic each year to ensure progress in a discipline within the space elevator project. The first such study was conducted in 2010 to evaluate the threat of space debris. The products from these studies are reports that are published to document progress in the development of space elevators. They can be downloaded at www.isec.org.
- International Cooperation – ISEC supports many activities around the globe to ensure that space elevators keep progressing towards a developmental program. International activities include coordinating with the two other major societies focusing on space elevators: the Japanese Space Elevator Association and EuroSpaceward. In addition, ISEC supports symposia and presentations at the International Academy of Astronautics and the International Astronautical Federation Congress each year.
- Publications – ISEC publishes a monthly e-Newsletter, its yearly study reports and an annual technical journal [CLIMB] to help spread information about space

elevators. In addition, there is a magazine filled with space elevator literature called Via Ad Astra.

- Reference material – ISEC is building a Space Elevator Library, including a reference database of Space Elevator related papers and publications. (see section before this on references)
- Outreach – People need to be made aware of the idea of a space elevator. Our outreach activity is responsible for providing the blueprint to reach societal, governmental, educational, and media institutions and expose them to the benefits of space elevators. ISEC members are readily available to speak at conferences and other public events in support of the space elevator. In addition to our monthly e-- Newsletter, we are also on Facebook, Linked In, and Twitter.
- Legal – The space elevator is going to break new legal ground. Existing space treaties may need to be amended. New treaties may be needed. International cooperation must be sought. Insurability will be a requirement. Legal activities encompass the legal environment of a space elevator - international maritime, air, and space law. Also, there will be interest within intellectual property, liability, and commerce law. Starting work on the legal foundation well in advance will result in a more rational product.
- History Committee – ISEC supports a small group of volunteers to document the history of space elevators. The committee's purpose is to provide insight into the progress being achieved currently and over the last century.
- Research Committee – ISEC is gathering the insight of researchers from around the world with respect to the future of space elevators. As scientific papers, reports and books are published, the research committee is pulling together this relative progress to assist academia and industry to progress towards an operational space elevator infrastructure.

ISEC is a traditional not-for-profit 501 (c) (3) organization with a board of directors and four officers: President, Vice President, Treasurer, and Secretary. inbox@isec.org / www.isec.org

Appendix B - Avoiding the Rocket Equation,

The Bottom Line: It is important to remember, Space Elevators are compatible and complementary to rocket architectures. The future needs both communities to work together. However, the first step is to help the rocket community understand the strengths of space elevators - "we can beat the rocket equation."

Problem: The Earth's gravity well has enabled life to develop. The atmosphere is dense, the movement of tectonic plates have enabled diversity and evolution, and the Moon supports us each day with a bright light and ocean tides. However, to escape, from the huge home base gravity well, has limited our migration off Earth. There is a lot to the gained from the defeat of adversaries. In this case, gravity. Our escape to the Moon for Apollo and the ability to pursue science in all parts of our solar system has demanded our best efforts and exceptional engineering feats. One over radius squared, from a big planet, is a difficult problem that needs to be defeated.

Examples: To escape is the first step of moving off planet (Low Earth orbit - achieved in 1957). It required the best of engineering feats by the Soviet Union. All space efforts since then have required huge masses of fuel and structure to leave the planet and gain orbital positions. This is usually explained in the terminology of gaining enough velocity to stay in orbit. To gain LEO, the accepted value is 9.3 km/sec velocity gained by burning fuel. Here lies the problem: We must burn fuel and send it out as exhaust to move the mass of the vehicle forward. Over the years, the consumption of 96% of the mass that starts on a launch pad is thrown away as the "cost of doing it this way." This included all the fuel needed to burn and push the rocket, the structures to hold the fuel, the rocket nozzles, and all the other structure needed to hold the payload safely in its grasp. We can all discuss the numbers, but a reasonable assumption is 4% of the mass on the pad gets to Low Earth Orbit. Another example was delivery of the Apollo lunar lander to the surface of the Moon was estimated at half of one percent of the launch mass (0.5%) reached the lunar surface.

Note: The reusability of stages and the cost effective approach that is being refined and exercised by all the rocket companies (lead by Blue Origin and SpaceX) are very good at being more efficient and even more cost effective - lowering the costs to orbit into a range of costs never expected. However, when one looks at the Tsiolkovsky rocket equation, one does not see any factor with reference to cost or reusability. As such, the mass to orbit is still not any more than the old numbers of 4% of pad launch mass. Goddard and Von Braun recognized this monumental problem and found ways to "work through it." An estimate of the SpaceX StarShip rocket capability is 100,000 kg to LEO with pad launch mass of 5,000,000 kg (estimates on wiki - 22 June 2020). This leads to only 2.0 % of launch pass mass to LEO. The system is very efficient and cost effective; however, it has not beaten the rocket equation and "big gravity."

The Rocket Equation[11]:   "The Tsiolkovsky rocket equation, classical rocket equation, or ideal rocket equation is a mathematical equation that describes the motion of vehicles that follow the basic principle of a rocket: a device that can apply acceleration to itself using thrust by expelling part of its mass with high velocity can thereby move due to the conservation of momentum. The equation relates the delta-v (the maximum change of velocity of the rocket if no other external forces act) to the effective exhaust velocity and the initial and final mass of a rocket, or other reaction engine. For any such maneuver (or journey involving a sequence of such maneuvers):"

$$\Delta v = v_e \ln \frac{m_0}{m_f} = I_{sp} g_0 \ln \frac{m_0}{m_f}$$

where:

Δv is delta-v – the maximum change of velocity of the vehicle (with no external forces acting).

m_0 is the initial total mass, including propellant, also known as wet mass.

m_f is the final total mass without propellant, also known as dry mass.

$v_e = I_{sp} g_0$ is the effective exhaust velocity, where:

I_{sp} is the specific impulse in dimension of time.

g_0 is standard gravity.

$\ln$ is the natural logarithm function.

The words of consequence are: "a device that can apply acceleration to itself using thrust by expelling part of its mass with high velocity can thereby move due to the conservation of momentum." The Tsiolkovsky rocket equation still responds to that critical factor called gravity. The Earth's gravity numbers have a consistent impact on effectiveness at liftoff and flight - DRACONIAN!

Space Elevator Strengths: For the GEO Region and beyond (including all solar system destinations) the Space Elevator "Beats the Rocket Equation." How is this done? Simple - it raises the cargo for each destination up to an altitude using electricity - not consuming rocket fuel and structure. As a result, the payload of the tether climber gains energy from the process. Using the Apex Anchor location as an indicator of the process, the payload has gained 100,000 km of potential energy and results in a horizontal velocity of 7.76 km/sec at release. This resulting energy gain has "enabled" the payload to go to Mars any day of the year (no waiting for 26 months for a launch window) and as rapidly as 61 days to Mars (over 200 releases across the planet's periodic dance of less than 200 days with many at 75-90 days to Mars). This is all achieved because the space elevator lifts the payload out of the gravity well and releases it when gravity is very low.

---

[11] Wikipedia 22 June 2020

This image shows the various release velocities at the lengths of space elevator tether. If one were to release at the Geosynchronous altitude, the payload would go into a geosynchronous orbit. As the height of release goes up, the velocity at release increases. The currently conceived length of a space elevator tether is 100,000 km; and, as such, provides enough velocity to reach the Moon in 10 hours or Mars in as little as 61 days. If one were to go to a 163,000 km altitude on a space elevator and release from the Apex Anchor, the payload could escape the Solar System without additional thrust (of course the mission would probably use gravity assist to gain velocity and rockets to correct the trajectory as it traversed open space).

Space Elevator
Launch Geometries[12]

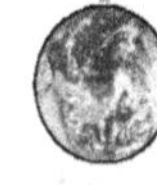

Conclusion #1: How can this be possible? Simple - a Space Elevator infrastructure need only defeat gravity and the traditional rocket equation once. Massive payloads to the Apex Anchor - raised by electricity - to be released at 7.76 km/sec towards destinations; daily, routinely, safely, and robustly all while being environmentally friendly.

Conclusion #2:   ISEC has shown that there must be major changes in the approach for humanity's migration off-planet. Some of these changes include:

Change of vision for interplanetary movement when delivery of mass is inexpensive, timely, environmentally friendly, daily, and supportive. It turns out the revelations in transportation capabilities of Space Elevators open up immense possibilities and ensures that humanity can "bring with them" the essential elements for survival and aggressive growth. This new vision of Space Elevator architectures will change the thinking for off-planet migration - We CAN bring it with us!

Movement off-planet will require complementary capabilities -both rocket portals and Space Elevator infrastructures - each with their own strengths and short-falls. Inserting payloads into Low Earth Orbits and moving people through the radiation belts rapidly are strengths of rockets while massive movement in a timely, routine, inexpensive and Earth friendly manner are the strengths of Space Elevators.

---

[12] Torla, James and Matthew Peet, "OPTIMIZATION OF LOW FUEL AND TIME-CRITICAL INTERPLANETARY TRANSFERS USING SPACE ELEVATOR APEX ANCHOR RELEASE: MARS, JUPITER AND SATURN," International Astronautics Congress (IAC-18-D4.3.4), Washington D.C., 2019.

A discussion of various destination mission needs when analyzing mass to location, will start the analysis of "how much carrying capability" is required by each supportive infrastructure: when, to where, and their priorities. In the past, the rocket approach valued light-weight and compact designs of support equipment while the Space Elevator permanent infrastructure will enable mass to be moved to desired destinations easily. The driving function for infrastructure design becomes a description of the customers' needs, not light weight designs.

An interesting insight in parallel with this analysis says that planetary scientists can be offered as much mass as they require for any of their missions. There will be zero restrictions for scientific instruments going to any place in the solar system - including the survival from the shake, rattle, and roll of rocket launches.. If you can not include it in one 14 metric ton payload capable tether climber, you can assemble parts at the Apex Anchor and release them once a day towards any destination.

After all the calculations and insight into the physics of lifting mass from the surface of the Earth, we are left with the conundrum of rockets. Can we continue to depend on a delivery system that only delivers a small percentage of mass to orbit?

Problem – Mass Deliver Percentage

Launch Vehicle	Mass on Pad (kg)	Mass Delivery	%
Apollo Saturn V	3,233,256	Lunar lander = 15,103	0.5
		ocean landing = 5,557	0.17
Atlas V	590,000	to GEO = 8,700	1.5
Falcon Heavy	1,420,788	to GEO = 26,700	1.9
Starship	4,000,000	to GEO = 21000	0.5
New Glenn	1,323,529	to GEO = 13,000	1

Space Elevators *answer the Conundrum of Rockets*

The conundrum of rockets is the simple realization that the delivery of mass to its destination is an insignificant percentage of the mass on the launch pad. The glaring example is the delivery of a half percent of the launch pad mass to the surface of the moon for Apollo 11. It is up to 2% for delivery to Geosynchronous Orbit and woefully small for delivery to Mars' orbit, much less Mars' surface. The question is why would you employ a methodology for delivery that only delivers less than one percent to your desired location (lets say the future Gateway around the Moon). The Space Elevator solves that conundrum by delivering 70% of the mass at liftoff (the other 30% is the tether climber and will be reused repeatedly) to GEO and beyond by leveraging electricity.

Appendix C - Body of Knowledge

Recently, a visitor to our International Space Elevator Consortium (ISEC) conference was quoted as saying, "You have a remarkable body of knowledge at www.isec.org. He was referring to the efforts of many scientists, engineers, and project/program professionals over the last 8 to 10 years. The leap in quality and currency shows that the Space Elevator is indeed twenty years beyond Dr. Edwards' breakthrough accomplishment saying "it can be done." What is amazing are the conclusions from this body of knowledge:
1.   Space Elevators are ready to initiate a developmental program
2.   The tether material has been produced in the laboratory for the needed strength (150 GPa) and continuous length (1 meter per minute production) (note; not both capabilities at once - yet). This 2D material will be ready for the development team.
3.   Space Elevators enable Missions off-planet with robust cargo movement as a complementary access to space with rockets.
4.   Space Elevators are environmentally friendly in operations and enable Space Based Solar Power to eliminate hundreds of coal burning plants.

ISEC is particularly proud of its latest year-long study entitled "Space Elevators are the Transportation Story of the 21st Century." This study report places Space Elevators into the near future and shows how they support critical missions. One such mission is the enabling of Space Based Solar Power. This mission will lead to a much cleaner global environment by eliminating hundreds (or thousands) of coal burning plants. The report also shows how to support Mars colonies and Lunar villages by supplying their cargo. In addition, this report illustrates research accomplished by ISEC with Arizona State University showing the strengths of Space Elevators for interplanetary missions. Can you imagine 61 days to Mars? How about daily departures to Mars (no 26 month wait)? In addition, Space Elevators enable a tremendous benefit with massive cargo movement (170,000 tonnes per year to GEO and beyond). All this is accomplished with the Space Elevator architecture as a complement to rockets. This Dual Space Access Architecture (rockets and Space Elevators) is complementary and compatible - not competitive.

Body of Knowledge - Current
The principal source for the following information is at www.isec.org.

A)   ISEC Studies: Latest engineering, management, operations, and developmental issues addressed in year-long studies conducted by Space Elevator experts. Download all 12 of these ISEC study reports in pdf for free at www.isec.org.
see the full list in the next appendix.

B)   International Academy of Astronautics Studies: In addition, there were three other major studies conducted on the modern Space Elevator; by the International Academy of Astronautics and the Obayashi Corporation.
2019 The Road to the Space Elevator Era - IAA International Academy of Astronautics (https://iaaspace.org) and 2014 Space Elevators: An Assessment of the Technological Feasibility and the Way Forward - IAA

C)   Obayashi Corporation: 2014 - The Space Elevator Construction Concept (https://www.obayashi.co.jp/en/news/detail/the_space_elevator_construction_concept.html)

D)   References and Citations are listed by major topic (over 750 titles available).

E)   Recently, the modern Space Elevator has been discussed within webinars that are accessible on ISEC website as well as YouTube. They are:
·   Oct 10, 2020 - Dual Space Access Architecture - Peter Swan (part of World Space Week activities - go to www.isec.org to sign up. (recently announced)
·   Aug 28, 2020 - Architectural Engineering for the Space Elevator - Michael Fitzgerald
·   July 17, 2020 - How Space Elevators Work: Physics Concepts - Dennis Wright
·   May 29, 2020 - Graphene: the Last Piece of the Space Elevator Puzzle? - Adrian Nixon
·   Apr 30, 2020 - Today's Space Elevator - Peter Swan

F)   A series of Architect's Notes address several of the significant choices along the path to development of a mega-project - website: isec.org

G)   Any questions can be forwarded to info@isec.org.

H)   Latest description of the Modern Space Elevator Architecture is shown in 10 videos on the ISEC youtube site. Check them out on the www.isec.org page to choose which one you would like to review.

Appendix D - Studies (ISEC, IAA, Obayashi)
List of International Space Elevator Study Reports Available
on www.isec.org or purchase from www.lulu.com

Year	Title
2022	Tether and Tether Climber interface assessment (in work)
2021	The Space Elevator is the Green Road to Space (this cocument)
2020	Galactic Harbour Interplanetary Mission Support
2020	Today's Space Elevator Assured Survivability Approach Space Debris
2019	Today's Space Elevator, Status as of Fall 2019
2018	Design Considerations for a Multi-Stage Space Elevator
2017	Design Considerations for a Software Space Elevator Simulator
2016	Design Considerations for Space Elevator Apex Anchor and GEO Node
2015	Design Considerations for Space Elevator Earth Port
2014	Space Elevator Architectures and Roadmaps
2013	Design Considerations for Space Elevator Tether Climbers
2012	Space Elevator Concept of Operations
2010	Space Elevator Survivability, Space Debris Mitigation
	and
2017	Space Elevator: A History

International Academy of Astronautics Studies (with participation from ISEC)

Year	Title
2019	The Road to the Space Elevator Era
2014	Space Elevators: An Assessment of the Technological Feasibility and the Way Forward
IAA	International Academy of Astronautics - sponsor of study www.iaaweb.org
go to:	Virginia Edition Publishing Company, Heinlein Prize Trust https://www.heinleinbooks.com/book-store

E. 1 Dangers from Nuclear Radiation (See also E. 2d & E. 2e)

Nuclear fission processes almost inevitably generate nuclei that are unstable,[fission products] . The resulting **radioactive decay**, (*alpha, beta* or *gamma*), poses **severe threats** to our **health** and our **environment**.

"All of us are exposed to radiation every day, from natural sources such as minerals in the ground, and man-made sources such as medical x-rays. According to the National Council on Radiation Protection and Measurements (NCRP), the average annual radiation dose per person in the U.S. is 6.2 millisieverts (620 <u>millirem</u>). The pie chart below shows the sources of this average dose", [U. S. EPA].

Figure 1 Sources of Exposure to Radiation, (U.S.A. Average at sea-level)

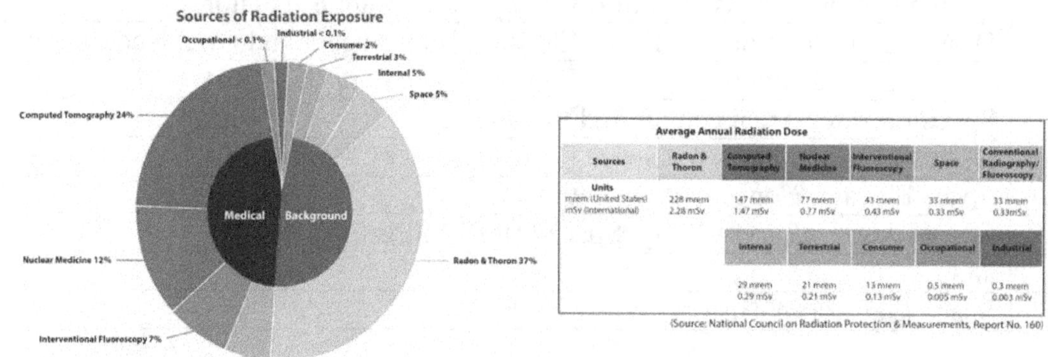

Average Annual Radiation Dose						
Sources	Radon & Thoron	Computed Tomography	Nuclear Medicine	Interventional Fluoroscopy	Space	Conventional Radiography/ Fluoroscopy
Units mrem (United States) mSv (International)	228 mrem 2.28 mSv	147 mrem 1.47 mSv	77 mrem 0.77 mSv	43 mrem 0.43 mSv	33 mrem 0.33 mSv	33 mrem 0.33mSv
		Internal	Terrestrial	Consumer	Occupational	Industrial
		29 mrem 0.29 mSv	21 mrem 0.21 mSv	13 mrem 0.13 mSv	0.5 mrem 0.005 mSv	0.3 mrem 0.003 mSv

(Source: National Council on Radiation Protection & Measurements, Report No. 160)

(See E. 2f Radiation in our Environment)

A radiation dose about **80 times greater than this**, (received over a **very short period of time**) has a **50% chance** of causing the individual to **die within 30-days,** [LD$_{50/30}$].
(see E. 3 Nuclear Weapons for further details.)

E. 2 Nuclear Electric Reactors

E. 2a History (1954 - 2018)

Figure 1 below, [History 1954], shows that the construction of new Reactors has decreased quite dramatically since the mid-1980's. The main reason is that originally Nuclear Electric Reactors were only licensed to operate for **40 years**. This means that the huge number of Reactors that were constructed from the 1960's through the 1980's should now be at or approaching their 'decommissioning' stage. Instead, because of upgrades to their 'operational control systems' and/ or improvements in their actual 'fuel-bundles' or Reactor design, their licenses have been extended to 60 years or even 80 years, [60 or 80 years]. This means that a Reactor that began operating in 1980 might well still be generating electrical energy in 2040 or 2050 or even 2060.

Figure 1 History of Nuclear Electric Reactors in the World

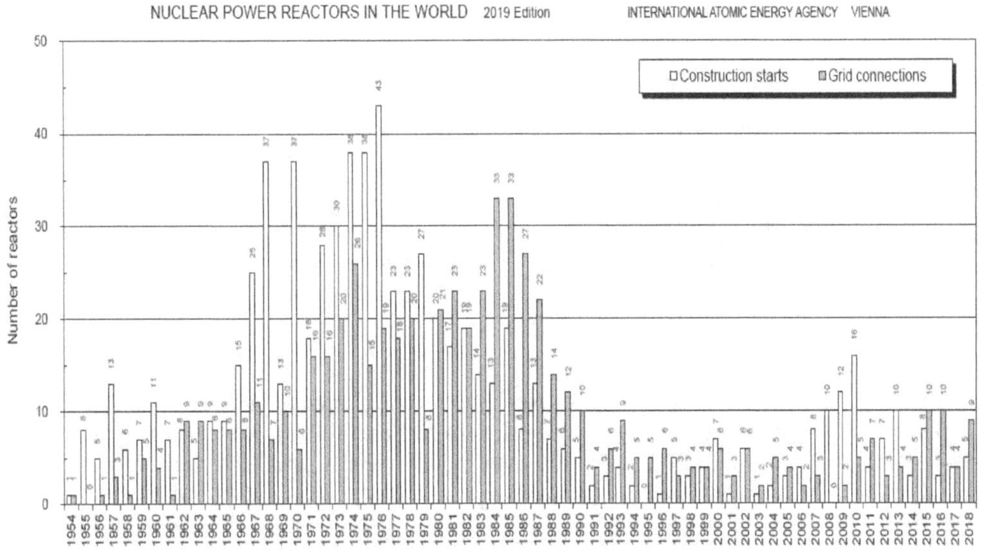

Figure 6. Annual construction starts and connections to the grid (1954 to 2018).

E. 2b Greenhouse-Gas Emissions from Fossil Fuels

Coal-Fired Plants generated about 40% of the World's electricity in 2018, while Oil and Natural Gas, (the 'other' Fossil Fuels), added about another 25%, [Electricity sources].

Figure 2 World Production of Electrical Energy

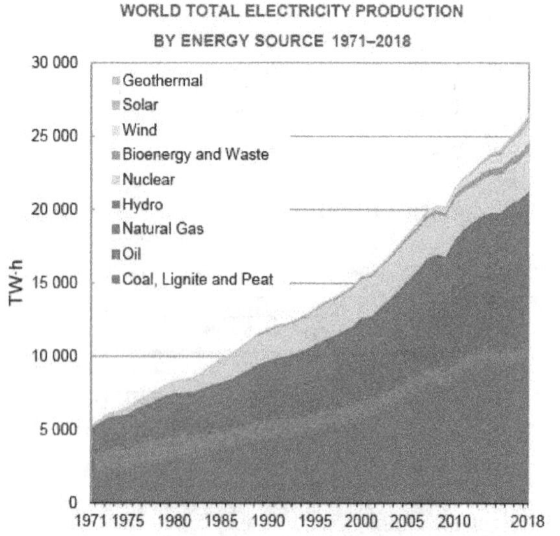

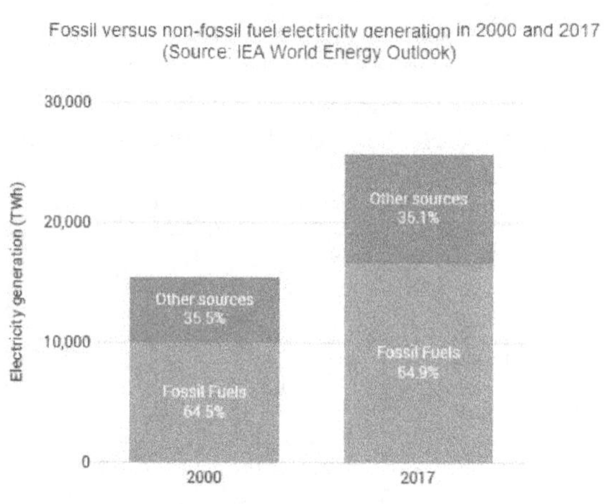

Note: Estimates are expressed in gross figures. Gross electricity production is the total electrical energy produced by all generating units and installations, including pumped storage, measured at the output terminals of the generators.

Comparing the emissions to the atmosphere for Nuclear electric plants and for just Coal-Fired electrical plants, (neglecting contributions from both Oil and Natural Gas), is startling. From EndCoal.org, [End Coal], as of mid-2020 there are over 6,600 operating coal-fired plants in the World, [Coal Units], with another 582 such plants in the Construction or 'permitted'-but-not-yet-started-construction stages. In 2019, their total CO_2 emissions exceeded **9 Billion tonnes,** [CO_2], (or **9 Gt – Giga-tonnes** - in metric units).

Shockingly, two separate Websites, (Our World in Data and IEA), [OWiD & IEA], claim that the actual World total of CO_2 emissions from coal is actually much **larger**, over **14 Billion tonnes**, and that the Annual World-Total of CO_2 emissions from **all sources** exceeds **33 Billion tonnes**.

World-wide efforts to reduce this through Carbon-Capture-and-Storage technologies, [CCS], are now increasing rapidly in their scope, but the projected CCS total of 40 Million tonnes, (~0.1 % of the World-emissions i 2020) implies it will take many years before this technology has a significant impact.

Admittedly, the nuclear **H-L-W** will be extremely disturbing until a safe and 'permanent' disposal method is devised. However, the huge volumes of carbon-dioxide emitted from nearly all coal-fired plants are routinely 'disposed of' by simple emission to the Earth's atmosphere.

E. 2c   Additional Nuclear Reactor Advantages

Figure 3 Energy Generation Fatalities OECD countries

"Accident fatalities from Energy generation", [Energy fatalities], is a topic we usually avoid thinking about, but, tragically, they do occur. The **large numbers** for the 'oldest' forms is the most shocking. That we seem to tolerate this may be because it has been part of our society for so many years, or even generations, that, sadly, it is seen as being 'inevitable'.

Perhaps we would be better served if we celebrate the relative safety of Nuclear, Solar and 'Onshore Wind' generation' **and** vehemently insist that the 'deadlier methods' must either enormously improve their safety records or cease operations!

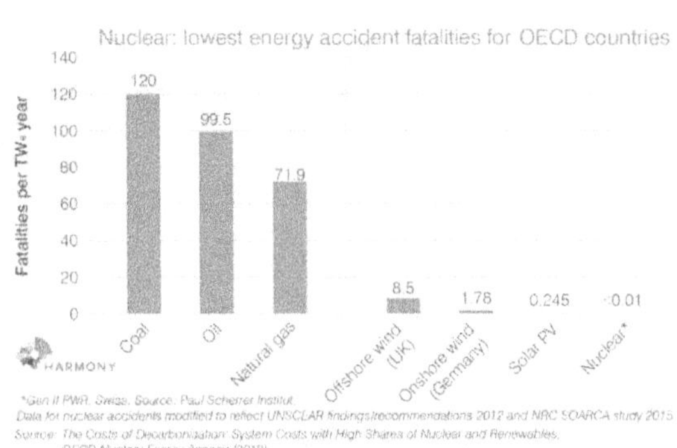

Figure E. 3b Land Required for Energy Generation

Another "seldom-mentioned" , but increasingly important further advantage of Nuclear is its 'tiny footprint', [footprints]. This is especially important when it is compared to the 'environmentally friendly' forms.

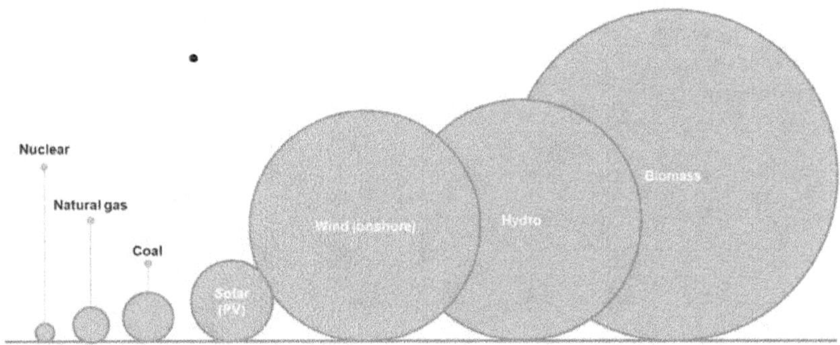

Land use (fuel mining and generating footprint) of electricity generation options per unit of electricity

E. 2d   Gen IV Reactors and High Level Waste, (H-L-W)

The American Nuclear Society, [ANS], ". . . believes that the development and deployment of advanced nuclear reactors based on fast-neutron fission technology is important to the sustainability, reliability, and security of the world's long-term energy supply. . . . extending by a hundred-fold the amount of energy extracted from the same amount of mined uranium. . . . Virtually all long-lived heavy elements are eliminated during fast reactor operation, leaving a small amount of **fission product waste** that requires assured isolation from the environment for **less than 500 years.**" (Note: **Emphasis added** by Ed.)

Figure 4 Spent Fuel Reprocessing

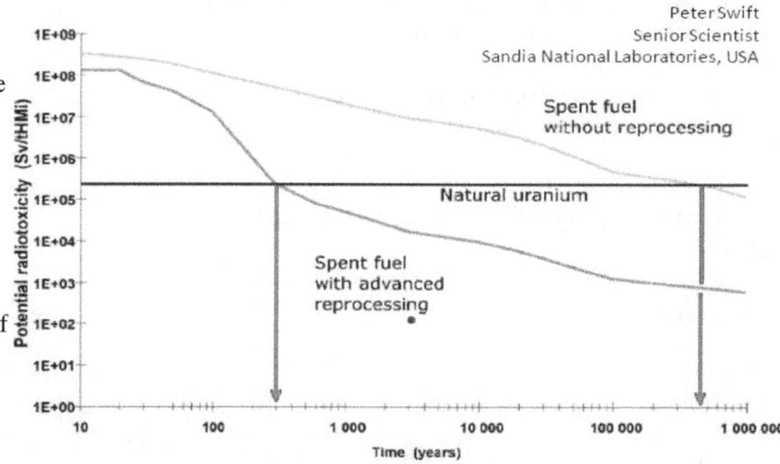

With current Reactors the 'spent fuel' is still **reusable** and will be even more-so with **Gen. IV Reactors**. With the fuel being reprocessed the 'quarantine period' drops to 'only' **several hundred years** instead of **several thousand Centuries !**

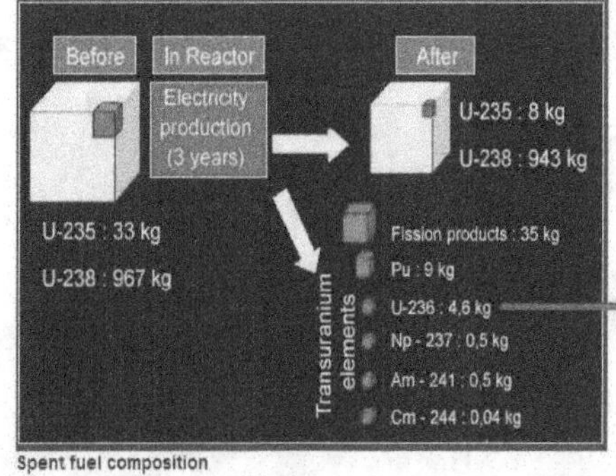

When the 'spent fuel', [spent fuel], is removed from the Reactor it still contains almost 25% of the original Uranium as well as an additional 27% of that mass in fissionable Plutonium 239. (There are also small amounts of the 'Actinides', Neptunium,(Np), Americium, (Am) and Curium, (Cm). These, when recovered, can be 'burned' in Gen. IV Reactors.) This 'new' fuel still has to be enriched in Uranium-235 before it can be re-utilized.

U-236 absorbs neutrons and is regarded as a 'poison' in Reactors. To compensate for this, the reprocessed fuel has to undergo an additional enrichment procedure.

E. 2e   Nuclear Electric Reactors -Reprocessing Fuel and H-L-W

When removed from the Reactor, (Main p 25), the spent-fuel is so radioactively 'hot' that special precautions have to be taken, [SKB][61].
{3 minute video at the site is excellent. B.M.}

Water-filled storage pools at the Central Interim Storage Facility for Spent Nuclear Fuel (CLAB) facility in Sweden. Storage ponds at reactors, and those at centralised facilities such as CLAB in Sweden, are 7-12 metres deep to allow for several metres of water over the used fuel (assembled in racks typically about 4 metres long and standing on end). The multiple racks are made of metal with neutron absorbers incorporated. The circulating water both shields and cools the fuel. These pools are robust constructions made of thick reinforced concrete with steel liners. Ponds at reactors are often designed to hold all the used fuel produced over the planned operating lifetime of the reactor.

https://www.skb.com/our-operations/clab/

Of the approximately **420 Kt** of 'spent fuel' that had been discharged from Nuclear Electric Reactors Worldwide, between 1954 and 2018, a total of **150 Kt** has been reprocessed, [SNF], (and the reprocessing **capacity** is approximately **5 Kt** per year). This reprocessing has led to ~**111,000 packages** of 'vitrified **H-L-W**', each of mass ~**493 Kg**, [vitrified]. As of 2015 ~**45 Kt** of the 'new' fuel remained in storage and **17 Kt** had been **recycled** into Reactors, [processing].   The only Nations that have active reprocessing plants, (as of 2018), [processing], are the U.K, France, Japan, Russia and India. They will be joined by China and South Korea during the next few years. The first 5 Nations above produced approximately 28% of the World Total Nuclear Electric Energy in 2019, [Energy 2019].

https://www.radioactivity.eu.com/site/pages/Vitrified_HA_Waste.htm

Container
Stainless steel
Height: 1 m 33.8 cm
Diameter: 43 cm
Mass: 92.5 kg

Package
Borosicilate glass
Volume: 175 litres
Waste mass: 400 kg
Radioactive products: 11 kg

Total Mass ≃ 493 Kg
Specific
Gravity ≃ 2.3

CDS-V vitrified waste packages
The quantity of waste vitrified at La Hague is forecast to be around 630 packages/year, based on an assumption of 10-15 packages per reactor for 850 tonnes of French fuel processed, plus 170 waste packages for international customers. This equates to 0.74 waste packages per tonne of spent fuel reprocessed. ©AREVA

Therefore, it is likely that, of the, (from above), **270 Kt** of 'spent fuel' that had **not** been reprocessed by 2018, perhaps 72%, (**~190 Kt**), will either **already** be in 'dry' storage or very soon **'headed there'**.

When reprocessing is planned, commercial Nuclear Electric companies seldom store their 'spent fuel' in 'on-site-ponds' for longer than 5 years because the pools are extremely expensive to construct and operate. Therefore we will 'guesstimate' that the **2018** total of **SNF** already in 'dry' storage or in the process of being 'repackaged' for dry storage is ~**160 Kt** . This is, of course, the actual mass of the **SNF**. It is still extremely radioactive and so requires extensive shielding.

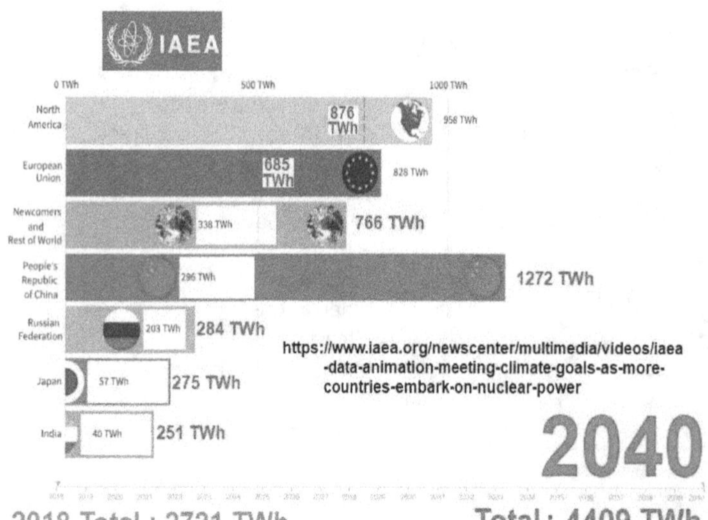

Nuclear generation of electric energy is predicted to increase another 60% from 2018 through 2040, [iaea 60%]. The two main factors, (as shown in the figure, left), are

from **China** planing an **increase** of more than 300%, and the **Newcomers etc.** more than doubling their Nuclear capacity. Japan, India and Russia also plan significant changes.

Only North America and the European Union will possibly **decrease** their Nuclear electric-energy generating capacity.

E. 2f Radiation in our Environment

Figure 5 Radiation Sources

A small piece of Uranium Ore that you could hold in your hand is about as radioactive as a large bunch of bananas, (which contain trace amounts of Potassium-40 - 'four-times **more** radioactive' than Uranium), [everyday objects], [Uranium half-life] . Other common foods and materials that are also radioactive: ". . . spinach, potatoes, oranges, brazil nuts, kitty litter, granite counter tops, the very air you breathe and even your own body!", [everyday objects]. If this seems outrageous it is because we tend to be totally unaware of the natural radioactivity that we encounter: from the Sun, from dirt or concrete that we walk on or build with, as well as the (**normally** extremely small) amount of **Radon gas** we constantly encounter, etc., etc.!

If we compare the above dangers to that from the 'dry casks',, (Ch. 4, **Section 4.5)**, you would have to **stand** at the boundary distance, (usually about **280 m** from the nearest casks), with at least 5 canisters in **clear view**, (i.e. not 'shielded' by other casks), for a **full year** to absorb a radiation dose of 20 mrem, [Dry Casks]. That would **increase** your **eventual cancer risk** by 0.004%, [Cancer Risk]. {Note from Ed. , moving about **ten times closer** to stand near the visible fence, would increase this life-time-cancer-risk by about 0.4%, or 1/250.}

E. 3 High Level Waste, (H-L-W) - Nuclear Weapons & Electric Reactors

E. 3a High-Level-Waste from Nuclear Weapons, (e.g. U.S.A. 1942 – 2040)
The U.S. has **four separate sites**, (in the **States** of New York, South Carolina, Washington and Idaho), that have been in the process of dealing with this 'nuclear weapons legacy' for the past **several decades**, and it is expected that the 'clean-up' will continue for at least another one or two decades, [USA NWW].

The unit of radioactivity, the **Curie**, [Curie]: **One Curie = 3.7×10¹⁰ decays per second,** (or 37 Billion decays per second). It was based on the measured, (in 1910), radiation emitted by **one gram of Ra²²⁶** , (Radium-226).

Figure 6 WVDP Canister Radioactivity, (in Curies)

Radioactivity per
Canister over Time

Decay Time after End of 1989 (yr)	Calculated Radioactivity per Canister (Ci)
0	109,600
1	106,900
2	104,300
5	97,080
10	86,230
100	10,730
1,000	63
10,000	13.2
100,000	5.2
1,000,000	3.1

Source: ORNL (1992a)

At the **West Valley Demonstration Project**,[WVDP], in New York State, 275 filled steel canisters of vitrified waste, each of approximate mass 2.1 **t,** were produced. The canisters are currently stored on-site in the shielded Chemical Process Cell. The radioactivity-level of each canister is shown in Figure 6, [WDVP].

Ironically, the purpose of the **WVDP** in New York State, was to **study the management of High-Level radioactive waste**. It operated only from 1966 to 1972, and was closed permanently in 1976. In 1977 it became a **legal requirement** for the **U.S.** to consider **all its 'spent fuel' as waste**, [Rossin 2014].

<u>**Savannah, S. C., River Site**</u>, (often referred to as **SRS**), by June 2019 had **4,190** filled canisters, [SRS double]⸴ accounting for about **9.0 Kt** of vitrified waste. Each cylindrical canister is **3.1 m tall, and 0.61 m in 'external diameter'** and **'canister + waste'** has a mass of almost **2.3 t**, [SRS canisters]. By the end of the process, (estimated around **2036**),the eventual total will be approximately **8200 canisters**, [SRS end].

Site #3, **Hanford, Wa.**, has an area of over 1,500 Km²), [History], much of it still contaminated from decades of the production of enriched Uranium and Plutonium for nuclear weapons. Despite decades of effort there is still, (as of 2019), a very large amount of 'clean-up' and mitigation to do. Currently, in dealing with the spent fuel-rods from the nuclear weapons era at Hanford, at least **400** cylindrical, steel **multi-canister-overpacks, (MCOs)**, have already been filled and safely stored in the **Canister Storage Building**, [CSB]. The overpacks are 64.3 cm in diameter, 432 cm in overall height, (sealed), and mass about 5.8 **t** each when filled and sealed, [MCO]. Since the **CSB** is designed to store **1,320** of the **MCOs**, (ibid.), an additional ∼ 900 **MCOs** can be **anticipated for the coming decade-or-so**. The entire clean-up may take until 2050 or even longer, [Completion].

There are also 34 smaller canisters, (mass ∼ 160 Kg each), of **H-L-W** stored at Hanford, (likely in the interim storage area), [German glass].

<u>**Idaho National Laboratory**</u> (INL), [Calcined]: In 2005 the total inventory of **SNF at INL** was **319 m³,**(approximately **2,800 t**) , in an **incredible 347 different types**, (claddings, history of use, levels of enrichment, coatings, reprocessed, etc.).

Also, another **~ 1,500 t** of **'calcined' H-L-W**, (**4,400 m³**), is in a total of **43** large steel containers stored in a total of six separate buildings at **INL** plus a concrete vault which has a design lifetime of 500 years.

In addition to the 'several-tonnes-each' types of canisters mentioned above, additional much larger dry-storage 'casks', (a steel-tube surrounded by thick concrete 'walls'), are used to isolate, for <u>**at least many decades**</u>, large quantities of **used-fuel-bundles** that have been removed from the reactors and are now likely destined for 'permanent-disposal'. There are at least eight such huge 'casks', each having a 'filled-mass' of **~100 t**, which includes **14 t** of fuel-bundles, , [INL SNF]

INL also has been responsible for processing H-L-W from the fleet of U. S. Nuclear-powered submarines, [USA subs]. This final shipment was the 118th such 100-tonne cask that was sent to the Naval Reactors Facility. Such a cask likely contains approximately 12 tonnes of spent nuclear fuel.

A significant amount of nuclear weapons radioactive-waste-processing remains to be done. This can be verified by scrolling through the many pages listing the progress of new programs at [Tank Waste].

The U.S. **Nuclear Weapons H-L-W** from 1942 to ~ the **mid-2030's**, (**Table 1** below) is:

Table 1

Storage Location	Projected Total *Vitrified & *SNF Waste (t)
Hanford, Washington	*32,000 + *8.5
Idaho National Laboratory, Idaho	54*1,500 + *2,800
Savannah River Site, South Carolina	*9,700 (2019) + *8,900 (2036)
West Valley Demonstration Project, N. Y.	*578
Hanford, Washington (Vitrification starts 2020's[49])	*41,650
INL, Idaho Naval Reactors Facility[56]	*36,000

Therefore, the **U.S.** will have, by the mid-2030's, more than **133 Kt** of **H-L-W**, (with at least **85 Kt** that already existed in packaged forms in **2018**).

Because FOI, (freedom-of-information-access), is quite advanced in the U.S., [FOIA], we shall **extrapolate** the above information to the other seven 'nuclear-weapons-Nations' based on the **maximum number of warheads** each had produced at any time between 1945 and 2014, [OWiD]. In doing this we are ignoring the common trend that almost anything we do gets more efficient as we gain experience ! This method could then lead to a **slight overestimate** of the amount of **H-L-W** produced in each case. The calculated results are shown below, (**Table 2**) :

Table 2

Name of Country	Max. # of 'war-heads'	Ratio to U.S.	Predicted H-L-W (Kt)
U.S.	23,368	1.000	('actual') 133
U.S.S.R. (Russia)	38,107	1.631	**217**
France	540	0.02311	**3.0**
U.K.	385	0.01648	**0.9**
PRC (China)	250	0.0107	**1.4**
India	110	0.00471	**0.6**
Pakistan	110	0.00471	**0.6**
Israel	**(est.)** 80	0.00342	**0.5**

Predicted 'World Total' from Nuclear Weapons in mid-2030's: 357 Kt
(Note: In Chapter 4 , (Section 4.2.5),, because the 'Hanford Vitrification Facility' is not yet operational, this estimate was reported as 315 Kt up to 2018.)

It is estimated, [IAEA-1587], that from **2018 to 2040** the discharge of 'spent, fuel' will likely be **~20 Kt** per year. The animation, [iaea 60%], predicts that Nuclear Electric generation may, by 2040, increase by 60%, but the much higher 'fuel burnup' and efficiency for generating electrical energy that the soon-to-be-starting-operations **Gen. IV Reactors** will almost certainly allow, [burnup], should largely compensate for that. Therefore, we predict, with some degree of confidence, that the 'spent fuel' discharge of **~20 Kt** per year will likely not change significantly for the next few decades.

Russia is nearing, (2024), operations for their earliest **Gen. IV Reactors** and plans to greatly

expand their number and 'type' plus aggressively reprocess **SNF**, [Russia Gen. IV].

China (**PRC**), [China expands], has **12** new Nuclear Electric Reactors under construction and **44** more Reactors in the planning stage, . It will likely become the largest "reprocessing-Nation" in the World. China will open a large reprocessing plant in 2020 and plans to **add another such plant each decade.**

Several Members of **GIF**, [G IV Members], are also approaching 'final' decisions about the specific types of Gen. IV systems they will pursue. Since Gen. IV Reactors greatly encourage the reprocessing of **SNF** while also 'burning' the Actinides in the fuel, it may be conceivable that the **H-L-W** problem will eliminated in the future, [No H-L-W].

Therefore, (keeping in mind the fact that up until 2050 Gen. IV Reactors probably will be only about 1/10 or 1/5 of the **operating** Electric Reactors), we are going to make the following, (perhaps 'crazy'), predictions :

Table 3

*(P 5 above, the 'spent-fuel' still contains ~96% 'wanted' elements. We used: 1 **t SNF =0.9 t 'new' fuel**)

Years	Total **SNF** **Discharged** (Kt)	% of SNF Reprocessed	'New' Fuel Produced* (Kt)	# of 'packages' Vitrified H-L-W for Disposal (0.5 t ea.)	'dry storage' SNF for Disposal ? (Kt)
2019 - 2030	220	35	69	**~51,000** + (52,000 Hanford)	1,500
2031 - 2050	380	50	171	**~141,000** + (32,000 Hanford)	2,000

Special Note: It was mentioned, (p 4 above), that at the **Hanford** 'Nuclear Weapons' site an additional **42 Kt** of vitrified **H-L-W** was expected to be produced, between about 2021 and 2036. For convenience this has simply been added, (prorated), to the Table above using the assumption that each 'package' will mass **~0.5 t.**

E. 4 Dismantling Nuclear Electric Reactors

Since most shut-downs of Reactors are planned well in advance, the future of the fuel-load is likely also determined well ahead of time so it will not be considered separately here.

In the USA, "To prepare for eventual decommissioning of a nuclear power plant, . . . the companies that operate them . . provide assurance that funds will be available . . through a trust fund . . **Decommissioning trust funds are <u>not</u> under the direct . . control of the generating companies, and use of the funds is limited to legitimate decommissioning expenses**." (*Emphasis added by Ed.),* [shut-down].

Similar requirements exist in Europe, and elsewhere, [Europe].

"Over 180 commercial, experimental or prototype reactors, over 500 research reactors, and several fuel cycle facilities have been retired from operation. Some of these have been fully dismantled. Most parts of a nuclear power plant do not become radioactive, or are contaminated at only very low levels. Most of the metal can be recycled. Proven techniques and equipment are available to dismantle nuclear facilities safely and these have now been well demonstrated in several parts of the world.
. . .   Decommissioning costs . . . for nuclear power plants are high relative to other industrial plants but are reducing, (and) *contribute only a small fraction of the total cost of electricity generation . . . The International Atomic Energy Agency (IAEA) has defined three options for decommissioning . . . Immediate Dismantling (or Early Site Release/'Decon' in the USA): . . . . Final dismantling or decontamination activities can begin within a few months . . . . the site is then available for re-use within a decade.*
Safe Enclosure ('Safstor') or deferred dismantling: This option postpones the final removal of controls for a longer period, usually in the order of 40 to 60 years . . .Th e facility is placed into a safe storage configuration until the eventual dismantling and decontamination activities occur after residual radioactivity has decayed.
Entombment (or 'Entomb'): . . . placing the facility into a condition that will allow the remaining on-site radioactive material to remain on-site without ever removing it totally. This option usually involves . . . encasing the facility in a long-lived structure such as concrete, that will last for a period of time to ensure the remaining radioactivity is no longer of concern." [Decommission].

E. 5 Yucca Mountain D.G.R., U.S.A. {- continued from Section 4.3)

Completing the 8 Km long main-access tunnel, (7.6 m diameter), plus 5 m exploratory side branches and various research alcoves cost 15 $Billion. Since 2014, the Nuclear Utilities have been receiving payment from a special Judgment Fund resulting from the failure to provide a promised Repository by 1998, [Yucca fiasco]. The payments may total $24 Billion by the mid-2020's but are not part of the Federal Budget so this has been ignored by Congress, (until at least September, 2020). Ongoing seismic activity in the region, [Tonopah quake], makes the future of the proposed D.G.R even more uncertain.

Congress chose the site in 1987 as the country's sole permanent nuclear repository.

Appendix to Ch. 4 FINAL References

[60 or 80 years] Office of Nuclear Energy 16 04 2020, What's the Lifespan for a Nuclear Reactor? Much Longer Than You Might Think https://www.energy.gov/ne/articles/whats-lifespan-nuclear-reactor-much-longer-you-might-think 02 05 2020

[ANS] American Nuclear Society **Fast Reactor Technology: A Path to Long-Term Energy** http://large.stanford.edu/courses/2013/ph241/roberts1/docs/ps74.pdf

[burnup] Nuclear Power, Fuel Burnup https://www.nuclear-power.net/nuclear-power/reactor-physics/reactor-operation/fuel-burnup/ 03 03 2020

and

Office of Nuclear Energy, 3 Advanced Reactor Systems to Watch by 2030
March 7, 2018 https://www.energy.gov/ne/articles/3-advanced-reactor-systems-watch-2030 14 12 2020

[Calcined]   U.S. NUCLEAR WASTE TECHNICAL REVIEW BOARD, CALCINED RADIOACTIVE WASTE, REVISION 2, JUNE 2020   calcined_hlw-rev2.pdf p 1-4

[Cancer Risk] International Energy Agency, IEA, **RADIATION, PEOPLE AND THE ENIRONMENT**, radiation0204.pdf , p 19

and

NIH PubMed Goldman, M. https://pubmed.ncbi.nlm.nih.gov/6761969/ Abstract 06 06 2020

[CCS] Green Car Congress, Mike Millikin 06 12 2020
https://www.greencarcongress.com/2020/12/20201206-ccs.html 07 12 2020

[China expands] World Nuclear Association, Nuclear Power in China, Nov. 2020, Further nuclear plants: operating, under construction and planned
https://www.world-nuclear.org/information-library/country-profiles/countries-a-f/china-nuclear-power.aspx 09 11 2020

and

World Nuclear Association, China's Nuclear Fuel Cycle, Nov. 2020, https://www.world-nuclear.org/information-library/country-profiles/countries-a-f/china-nuclear-fuel-cycle.aspx 09 11 2020

and

World Nuclear News, Chinese reprocessing plant to start up in 2030, 24 September 2015 https://www.world-nuclear-news.org/WR-Chinese-reprocessing-plant-to-start-up-in-2030-2409155.html 09 12 2020

and

Zhang, Hui Bulletin of the Atomic Scientists, Pinpointing China's new plutonium reprocessing plant, May 5, 2020 https://thebulletin.org/2020/05/pinpointing-chinas-new-plutonium-reprocessing-plant/ 09 12 2020

[CO$_2$] Coal Plants by Region: Annual CO2
https://docs.google.com/spreadsheets/d/1Kn2Hs8SwIAwAt5V7iZUUgAjKs3GfeEncKCHmor7kbkg /edit#gid=0 08 09 2020

[Coal Units] EndCoal.org https://docs.google.com/spreadsheets/d/1JKJJa-jwK6YpkEQKP2bcENHR2yoS40ur8baQnIXHtIU/edit#gid=581593862 08 09 2020

[Completion] HANFORD SITE Cleanup Completion Framework, Jan. 2013 Hanford Comp_Framework_Jan_1-23-13-lfm.pdf p 7

[CSB] Hanford Site, Canister Storage Building and Interim Storage Area https://www.hanford.gov/page.cfm/CSB 03 04 2020

[Curie] U.S. Department of Health & Human Services, REMM,
https://www.remm.nlm.gov/radmeasurement.htm 06 06 2020

[Decommission] World Nuclear Association, Decommissioning Nuclear Facilities, Aug. 2020
https://www.world-nuclear.org/information-library/nuclear-fuel-cycle/nuclear-wastes/decommissioning-nuclear-facilities.aspx 16 04 2020

[Dry Casks] Al-Othmany, Dheya Shujaa **Radiation Shielding Analysis for a Spent Fuel Storage Cask under Normal Storage Conditions,** Advances in Physics Theories and Applications, Vol.79, 2019 49416-53068-1-PB.pdf, Abstract p 1

[Electricity sources] International Atomic Energy Agency, Nuclear Power Reactors in the World, 2019 19-00521_web.pdf p15

[End Coal] Main Site: https://endcoal.org/about/ , and
https://endcoal.org/global-coal-plant-tracker/summary-statistics/ 08 09 2020

[Energy fatalities] World Nuclear Association Image Library https://www.world-nuclear.org/gallery/the-harmony-programme/energy-accident-fatalities-for-oecd-countries.aspx 14 04 2

[Energy 2019] World Nuclear Association, Nuclear share figures, 2009-2019, (May 2020),
https://www.world-nuclear.org/information-library/facts-and-figures/nuclear-generation-by-country.aspx 11 05 2020 (Note: % calculated by Ed.)

[Europe] European Commission, Decommissioning of nuclear facilities,
https://ec.europa.eu/energy/topics/nuclear-energy/decommissioning-nuclear-facilities_en 05 28 2020

[everyday objects] Schwarcz, Joe 15 Mar 2018 McGill Office for Science and Society, "Is it true that bananas are radioactive?" https://www.mcgill.ca/oss/article/you-asked/it-true-banana-radioactive 11 04 202

[fission products] https://www.radioactivity.eu.com/site/pages/Fission_Products.htm 19 03 2020

[FOIA] History, Freedom of Information Act, Aug 21, 2018,
https://www.history.com/topics/1960s/freedom-of-information-act#:~:text=The %20Freedom%20of%20Information%20Act%2C%20or%20FOIA%2C%20was,right%20to %20access%20records%20from%20any%20federal%20agency 17 09 2020

[footprints] World Nuclear Association Nuclear Energy and Sustainable Development 04 2020
https://www.world-nuclear.org/information-library/energy-and-the-environment/nuclear-energy-and-sustainable-development.aspx 19 05 2020

[German glass] U.S. NUCLEAR WASTE TECHNICAL REVIEW BOARD VITRIFIED HIGH-LEVEL RADIOACTIVE WASTE, vitrified_hlw.pdf pp 2-4

[G IV Members] GEN IV International Forum, GIF Membership, https://www.gen-4.org/gif/jcms/c_9492/members 09 01 2020

[History] Hanford Site, Hanford History https://www.hanford.gov/page.cfm/HanfordHistory 04 04 2020

[History 1954] IAEA 2019, Nuclear Power Reactors in the World, RDS-2-39_web.pdf p 80

[IAEA-1587] International Atomic Energy Agency, Spent Fuel Reprocessing Options, Aug., 2008 TE_1587_web.pdf, p 2

[iaea 60%] International Atomic Energy Agency https://www.iaea.org/newscenter/multimedia/videos/ iaea-data-animation-meeting-climate-goals-as-more-countries-embark-on-nuclear-power 09 05 2020

[INL SNF] Office of Environmental Management, Spent Nuclear Fuel, https://www.energy.gov/em/services/waste-management/nuclear-materials-disposition/ spent-nuclear-fuel 26 07 2020

[LD$_{50/30}$] Nuclear Power, https://www.nuclear-power.net/nuclear-engineering/radiation-protection/ radiobiology/deterministic-effects/lethal-dose-of-radiation/ 17 09 2020

[MCO] United States Department of Energy Office of Environmental Management, N Reactor (U-metal) Fuel Characteristics for Disposal Criticality Analysis, May 2000 ML14127A144.pdf pp 23 - 34

[No H-L-W] Gen IV Nuclear Energy Systems Services, LLC, Exploring the Imperative of Metamorphosization of America's Electricity Generation, Space Exploration, Environmental Stewardship, & more..., https://www.genivnuclearenergysystems.com/home.html 09 06 2020

[OWiD] Our World in Data, Hannah Ritchie and Max Roser, Stockpiles of nuclear weapons, (no date given), https://ourworldindata.org/nuclear-weapons 04 28 2020 & 06 04 2020

[OWiD & IEA] Our World in Data, Hannah Ritchie and Max Roser, **CO2 emissions by fuel**, (no date given) https://ourworldindata.org/emissions-by-fuel 01 12 2020

and

International Energy Agency, IEA, 11 02 2020 https://www.iea.org/articles/global-co2-emissions-in-2019 01 12 2020

[processing] World Nuclear Association, Processing of Used Nuclear Fuel, (June, 2018), https://www.world-nuclear.org/information-library/nuclear-fuel-cycle/fuel-recycling/ processing-of-used-nuclear-fuel.aspx 05 03 2020

[Rossin 2014] Rossin, A. David U.S. Policy on Spent Fuel Reprocessing: The Issues,   **Frontline**: Why Do Americans Fear Nuclear Power ? **3. The Carter Policy**, https://www.pbs.org/wgbh/pages/frontline/shows/reaction/readings/rossin.html 06 06 2020

[Russia Gen. IV] World Nuclear Association, Nuclear Power in Russia, Nov. 2020 https://www.world-nuclear.org/information-library/country-profiles/countries-o-s/russia-nuclear-power.aspx 04 06 2020

[shut-down] Nuclear Energy Institute, Decommissioning Nuclear Power Plants August 2016,
https://nei.org/resources/fact-sheets/decommissioning-nuclear-power-plants#:~:text=Decommissioning%20is%20the%20process%20by%20which%20nuclear%20power,licenses%20granted%20by%20the%20U.S.%20Nuclear%20Regulatory%20Commission 04 03 2020

      and

World Nuclear Association, Decommissioning Nuclear Facilities , August 2020
https://www.world-nuclear.org/information-library/nuclear-fuel-cycle/nuclear-wastes/decommissioning-nuclear-facilities.aspx 07 03 2020

[SKB] Clab – Central Interim Storage Facility for Spent Nuclear Fuel, https://www.skb.com/our-operations/clab/ 18 04 2020

[SNF] International Energy Agency, IAEA-TECDOC-1587   TE_1587_web.pdf, p 2

[spent fuel] Radioactivity.EU.com, Spent fuel composition
https://www.radioactivity.eu.com/site/pages/Spent_Fuel_Composition.htm   03 08 2020

[SRS canisters] Folga, S .M. *et al* High-Level Waste Inventory, Characteristics, Generation, ANI7EAD/TM-17, 1996, 30001802.pdf p 20

[SRS double] Project to Double-Stack SRS Waste Canisters Gains Ground, Sept. 2020
https://www.energy.gov/em/articles/project-double-stack-srs-waste-canisters-gains-ground 07 10 2020

[SRS end]   SRS Fact Sheet, August, 2019 Defense Waste Processing Facility.pdf

[Tank Waste] Office of Environmental Management, Tank Waste
https://www.energy.gov/em/listings/tank-waste 27 11 2020

[Tonopah quake] Sadler, John May 19, 2020, Las Vegas Sun, Tonopah quake bolsters Nevada's case against Yucca dumpsite, https://lasvegassun.com/news/2020/may/19/tonopah-quake-bolsters-nevadas-case-against-yucca/ 22 05 2020

[Uranium half-life] Reference.com, What Is the Half-Life of Uranium? 25 03 2020
https://www.reference.com/science/half-life-uranium-d4cc5262a78f7acd 11 05 2020

[U. S. EPA] U. S. Environmental Protection Agency, https://www.epa.gov/radiation/radiation-sources-and-doses 16 09 2030

[USA NWW] Folga, S .M. *et al* High-Level Waste Inventory, Characteristics, Generation, . . . ANI7EAD/TM-17, 1996, 30001802.pdf ppg. 1 - 99

[USA subs] Office of Environmental Management, Workers Complete Fuel Transfer Work Scope at Idaho Site, March 5, 2019 https://www.energy.gov/em/articles/workers-complete-fuel-transfer-work-scope-idaho-site 12 09 2020

[vitrified] Radioactivity.EU.com, Vitrified High Level Waste
https://www.radioactivity.eu.com/site/pages/Vitrified_HA_Waste.htm 03 04 2020

[WVDP] Folga, S .M. *et al* High-Level Waste Inventory, Characteristics, Generation, . . .
ANI7EAD/TM-17, 1996, 30001802.pdf ppg. 92 - 99, Radioactivity chart p 42

and

WEST VALLEY DEMONSTRATION PROJECT ANNUAL SITE ENVIRONMENTAL REPORT CALENDAR
YEAR 2002, ENVIRONMENTAL PROGRAM INFORMATION,
814963.pdf, Chapter 1, p 1 - 8

[Yucca fiasco] DIXON, DARIUS 11/30/2013 Policico Magazine,
https://www.politico.com/story/2013/11/nuclear-waste-fiasco-100450 07 05 2020

Appendix F - Space Solar Power Backup

A - Comparison of SSP to terrestrial solar efficiency, measured in MW/km2, for terrestrial solar power plants (blue) and SPS systems (magenta). For terrestrial solar power plants, the known or modeled power output (MWAC) is divided by the total site area of the plant. For SPS systems, the power proposed to be delivered is divided by the proposed area of the rectenna.

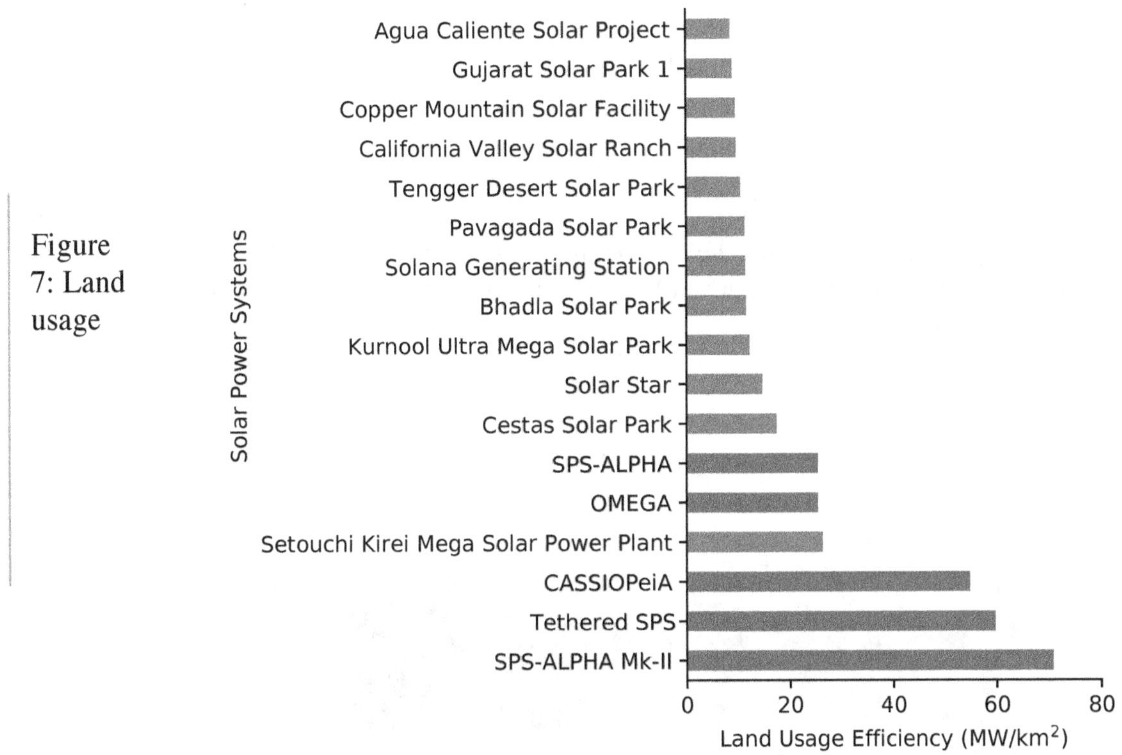

Figure 7: Land usage

B - SSP deployment via rocket
Given the SPS systems in Table 3, we examine the number of launches required for the deployment of a single system instance to GEO. This is shown graphically in Figure 8.

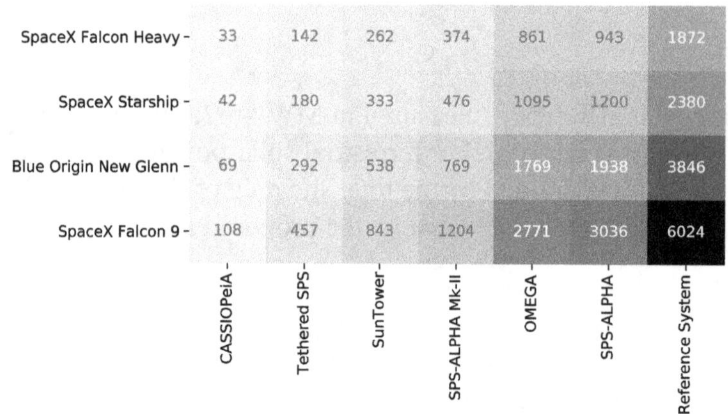

Figure 8: Number of rocket launches required to deploy a single SPS instance of each type using a variety of rocket launch systems. Rockets are ordered bottom-to-top by increasing maximum payload mass to GTO. SPS systems are ordered left-to-right by increasing hardware mass.

C - SSP deployment via space elevator

Given the SPS systems in Table 3, we examine the number of lifts on each SE system type required for the deployment of a single SPS system instance to GEO. This is shown graphically in Figure 9.

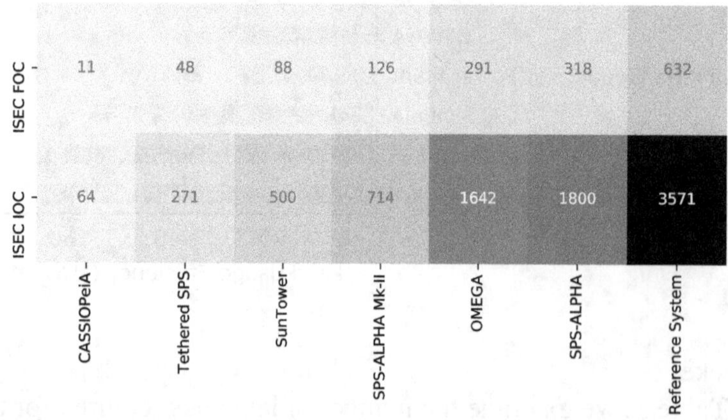

Figure 9: Number of lifts required to deploy a single SPS instance of each type using a single space elevator (SE). Lifts by climbers on each SE system are expected to occur at a frequency of once per day. SEs are ordered bottom-to-top by increasing throughput to GEO. SPS systems are ordered left-to-right by increasing hardware mass.

www.ingramcontent.com/pod-product-compliance
Lightning Source LLC
Chambersburg PA
CBHW081306250726
48662CB00008B/2415